The Sun: A Very Short Introduction

VERY SHORT INTRODUCTIONS are for anyone wanting a stimulating and accessible way into a new subject. They are written by experts, and have been translated into more than 45 different languages.

The series began in 1995, and now covers a wide variety of topics in every discipline. The VSI library currently contains over 650 volumes—a Very Short Introduction to everything from Psychology and Philosophy of Science to American History and Relativity—and continues to grow in every subject area.

Very Short Introductions available now:

ABOLITIONISM Richard S. Newman
THE ABRAHAMIC RELIGIONS Charles L. Cohen
ACCOUNTING Christopher Nobes
ADAM SMITH Christopher J. Berry
ADOLESCENCE Peter K. Smith
ADVERTISING Winston Fletcher
AERIAL WARFARE Frank Ledwidge
AESTHETICS Bence Nanay
AFRICAN AMERICAN RELIGION Eddie S. Glaude Jr
AFRICAN HISTORY John Parker and Richard Rathbone
AFRICAN POLITICS Ian Taylor
AFRICAN RELIGIONS Jacob K. Olupona
AGEING Nancy A. Pachana
AGNOSTICISM Robin Le Poidevin
AGRICULTURE Paul Brassley and Richard Soffe
ALBERT CAMUS Oliver Gloag
ALEXANDER THE GREAT Hugh Bowden
ALGEBRA Peter M. Higgins
AMERICAN CULTURAL HISTORY Eric Avila
AMERICAN FOREIGN RELATIONS Andrew Preston
AMERICAN HISTORY Paul S. Boyer
AMERICAN IMMIGRATION David A. Gerber
AMERICAN LEGAL HISTORY G. Edward White
AMERICAN NAVAL HISTORY Craig L. Symonds
AMERICAN POLITICAL HISTORY Donald Critchlow
AMERICAN POLITICAL PARTIES AND ELECTIONS L. Sandy Maisel
AMERICAN POLITICS Richard M. Valelly
THE AMERICAN PRESIDENCY Charles O. Jones
THE AMERICAN REVOLUTION Robert J. Allison
AMERICAN SLAVERY Heather Andrea Williams
THE AMERICAN WEST Stephen Aron
AMERICAN WOMEN'S HISTORY Susan Ware
ANAESTHESIA Aidan O'Donnell
ANALYTIC PHILOSOPHY Michael Beaney
ANARCHISM Colin Ward
ANCIENT ASSYRIA Karen Radner
ANCIENT EGYPT Ian Shaw
ANCIENT EGYPTIAN ART AND ARCHITECTURE Christina Riggs
ANCIENT GREECE Paul Cartledge
THE ANCIENT NEAR EAST Amanda H. Podany
ANCIENT PHILOSOPHY Julia Annas
ANCIENT WARFARE Harry Sidebottom
ANGELS David Albert Jones
ANGLICANISM Mark Chapman
THE ANGLO-SAXON AGE John Blair
ANIMAL BEHAVIOUR Tristram D. Wyatt

THE ANIMAL KINGDOM Peter Holland
ANIMAL RIGHTS David DeGrazia
THE ANTARCTIC Klaus Dodds
ANTHROPOCENE Erle C. Ellis
ANTISEMITISM Steven Beller
ANXIETY Daniel Freeman and Jason Freeman
THE APOCRYPHAL GOSPELS Paul Foster
APPLIED MATHEMATICS Alain Goriely
ARCHAEOLOGY Paul Bahn
ARCHITECTURE Andrew Ballantyne
ARISTOCRACY William Doyle
ARISTOTLE Jonathan Barnes
ART HISTORY Dana Arnold
ART THEORY Cynthia Freeland
ARTIFICIAL INTELLIGENCE Margaret A. Boden
ASIAN AMERICAN HISTORY Madeline Y. Hsu
ASTROBIOLOGY David C. Catling
ASTROPHYSICS James Binney
ATHEISM Julian Baggini
THE ATMOSPHERE Paul I. Palmer
AUGUSTINE Henry Chadwick
AUSTRALIA Kenneth Morgan
AUTISM Uta Frith
AUTOBIOGRAPHY Laura Marcus
THE AVANT GARDE David Cottington
THE AZTECS David Carrasco
BABYLONIA Trevor Bryce
BACTERIA Sebastian G. B. Amyes
BANKING John Goddard and John O. S. Wilson
BARTHES Jonathan Culler
THE BEATS David Sterritt
BEAUTY Roger Scruton
BEHAVIOURAL ECONOMICS Michelle Baddeley
BESTSELLERS John Sutherland
THE BIBLE John Riches
BIBLICAL ARCHAEOLOGY Eric H. Cline
BIG DATA Dawn E. Holmes
BIOGRAPHY Hermione Lee
BIOMETRICS Michael Fairhurst
BLACK HOLES Katherine Blundell
BLOOD Chris Cooper
THE BLUES Elijah Wald
THE BODY Chris Shilling
THE BOOK OF COMMON PRAYER Brian Cummings
THE BOOK OF MORMON Terryl Givens
BORDERS Alexander C. Diener and Joshua Hagen
THE BRAIN Michael O'Shea
BRANDING Robert Jones
THE BRICS Andrew F. Cooper
THE BRITISH CONSTITUTION Martin Loughlin
THE BRITISH EMPIRE Ashley Jackson
BRITISH POLITICS Anthony Wright
BUDDHA Michael Carrithers
BUDDHISM Damien Keown
BUDDHIST ETHICS Damien Keown
BYZANTIUM Peter Sarris
C. S. LEWIS James Como
CALVINISM Jon Balserak
CANCER Nicholas James
CAPITALISM James Fulcher
CATHOLICISM Gerald O'Collins
CAUSATION Stephen Mumford and Rani Lill Anjum
THE CELL Terence Allen and Graham Cowling
THE CELTS Barry Cunliffe
CHAOS Leonard Smith
CHARLES DICKENS Jenny Hartley
CHEMISTRY Peter Atkins
CHILD PSYCHOLOGY Usha Goswami
CHILDREN'S LITERATURE Kimberley Reynolds
CHINESE LITERATURE Sabina Knight
CHOICE THEORY Michael Allingham
CHRISTIAN ART Beth Williamson
CHRISTIAN ETHICS D. Stephen Long
CHRISTIANITY Linda Woodhead
CIRCADIAN RHYTHMS Russell Foster and Leon Kreitzman
CITIZENSHIP Richard Bellamy
CIVIL ENGINEERING David Muir Wood
CLASSICAL LITERATURE William Allan
CLASSICAL MYTHOLOGY Helen Morales
CLASSICS Mary Beard and John Henderson
CLAUSEWITZ Michael Howard
CLIMATE Mark Maslin

CLIMATE CHANGE Mark Maslin
CLINICAL PSYCHOLOGY Susan Llewelyn and Katie Aafjes-van Doorn
COGNITIVE NEUROSCIENCE Richard Passingham
THE COLD WAR Robert McMahon
COLONIAL AMERICA Alan Taylor
COLONIAL LATIN AMERICAN LITERATURE Rolena Adorno
COMBINATORICS Robin Wilson
COMEDY Matthew Bevis
COMMUNISM Leslie Holmes
COMPARATIVE LITERATURE Ben Hutchinson
COMPLEXITY John H. Holland
THE COMPUTER Darrel Ince
COMPUTER SCIENCE Subrata Dasgupta
CONCENTRATION CAMPS Dan Stone
CONFUCIANISM Daniel K. Gardner
THE CONQUISTADORS Matthew Restall and Felipe Fernández-Armesto
CONSCIENCE Paul Strohm
CONSCIOUSNESS Susan Blackmore
CONTEMPORARY ART Julian Stallabrass
CONTEMPORARY FICTION Robert Eaglestone
CONTINENTAL PHILOSOPHY Simon Critchley
COPERNICUS Owen Gingerich
CORAL REEFS Charles Sheppard
CORPORATE SOCIAL RESPONSIBILITY Jeremy Moon
CORRUPTION Leslie Holmes
COSMOLOGY Peter Coles
COUNTRY MUSIC Richard Carlin
CRIME FICTION Richard Bradford
CRIMINAL JUSTICE Julian V. Roberts
CRIMINOLOGY Tim Newburn
CRITICAL THEORY Stephen Eric Bronner
THE CRUSADES Christopher Tyerman
CRYPTOGRAPHY Fred Piper and Sean Murphy
CRYSTALLOGRAPHY A. M. Glazer
THE CULTURAL REVOLUTION Richard Curt Kraus
DADA AND SURREALISM David Hopkins
DANTE Peter Hainsworth and David Robey
DARWIN Jonathan Howard
THE DEAD SEA SCROLLS Timothy H. Lim
DECADENCE David Weir
DECOLONIZATION Dane Kennedy
DEMOCRACY Bernard Crick
DEMOGRAPHY Sarah Harper
DEPRESSION Jan Scott and Mary Jane Tacchi
DERRIDA Simon Glendinning
DESCARTES Tom Sorell
DESERTS Nick Middleton
DESIGN John Heskett
DEVELOPMENT Ian Goldin
DEVELOPMENTAL BIOLOGY Lewis Wolpert
THE DEVIL Darren Oldridge
DIASPORA Kevin Kenny
DICTIONARIES Lynda Mugglestone
DINOSAURS David Norman
DIPLOMACY Joseph M. Siracusa
DOCUMENTARY FILM Patricia Aufderheide
DREAMING J. Allan Hobson
DRUGS Les Iversen
DRUIDS Barry Cunliffe
DYNASTY Jeroen Duindam
DYSLEXIA Margaret J. Snowling
EARLY MUSIC Thomas Forrest Kelly
THE EARTH Martin Redfern
EARTH SYSTEM SCIENCE Tim Lenton
ECONOMICS Partha Dasgupta
EDUCATION Gary Thomas
EGYPTIAN MYTH Geraldine Pinch
EIGHTEENTH-CENTURY BRITAIN Paul Langford
THE ELEMENTS Philip Ball
EMOTION Dylan Evans
EMPIRE Stephen Howe
ENERGY SYSTEMS Nick Jenkins
ENGELS Terrell Carver
ENGINEERING David Blockley
THE ENGLISH LANGUAGE Simon Horobin
ENGLISH LITERATURE Jonathan Bate

THE ENLIGHTENMENT John Robertson
ENTREPRENEURSHIP Paul Westhead and Mike Wright
ENVIRONMENTAL ECONOMICS Stephen Smith
ENVIRONMENTAL ETHICS Robin Attfield
ENVIRONMENTAL LAW Elizabeth Fisher
ENVIRONMENTAL POLITICS Andrew Dobson
EPICUREANISM Catherine Wilson
EPIDEMIOLOGY Rodolfo Saracci
ETHICS Simon Blackburn
ETHNOMUSICOLOGY Timothy Rice
THE ETRUSCANS Christopher Smith
EUGENICS Philippa Levine
THE EUROPEAN UNION Simon Usherwood and John Pinder
EUROPEAN UNION LAW Anthony Arnull
EVOLUTION Brian and Deborah Charlesworth
EXISTENTIALISM Thomas Flynn
EXPLORATION Stewart A. Weaver
EXTINCTION Paul B. Wignall
THE EYE Michael Land
FAIRY TALE Marina Warner
FAMILY LAW Jonathan Herring
FASCISM Kevin Passmore
FASHION Rebecca Arnold
FEDERALISM Mark J. Rozell and Clyde Wilcox
FEMINISM Margaret Walters
FILM Michael Wood
FILM MUSIC Kathryn Kalinak
FILM NOIR James Naremore
THE FIRST WORLD WAR Michael Howard
FOLK MUSIC Mark Slobin
FOOD John Krebs
FORENSIC PSYCHOLOGY David Canter
FORENSIC SCIENCE Jim Fraser
FORESTS Jaboury Ghazoul
FOSSILS Keith Thomson
FOUCAULT Gary Gutting
THE FOUNDING FATHERS R. B. Bernstein
FRACTALS Kenneth Falconer
FREE SPEECH Nigel Warburton
FREE WILL Thomas Pink
FREEMASONRY Andreas Önnerfors
FRENCH LITERATURE John D. Lyons
THE FRENCH REVOLUTION William Doyle
FREUD Anthony Storr
FUNDAMENTALISM Malise Ruthven
FUNGI Nicholas P. Money
THE FUTURE Jennifer M. Gidley
GALAXIES John Gribbin
GALILEO Stillman Drake
GAME THEORY Ken Binmore
GANDHI Bhikhu Parekh
GARDEN HISTORY Gordon Campbell
GENES Jonathan Slack
GENIUS Andrew Robinson
GENOMICS John Archibald
GEOFFREY CHAUCER David Wallace
GEOGRAPHY John Matthews and David Herbert
GEOLOGY Jan Zalasiewicz
GEOPHYSICS William Lowrie
GEOPOLITICS Klaus Dodds
GERMAN LITERATURE Nicholas Boyle
GERMAN PHILOSOPHY Andrew Bowie
GLACIATION David J. A. Evans
GLOBAL CATASTROPHES Bill McGuire
GLOBAL ECONOMIC HISTORY Robert C. Allen
GLOBALIZATION Manfred Steger
GOD John Bowker
GOETHE Ritchie Robertson
THE GOTHIC Nick Groom
GOVERNANCE Mark Bevir
GRAVITY Timothy Clifton
THE GREAT DEPRESSION AND THE NEW DEAL Eric Rauchway
HABERMAS James Gordon Finlayson
THE HABSBURG EMPIRE Martyn Rady
HAPPINESS Daniel M. Haybron
THE HARLEM RENAISSANCE Cheryl A. Wall
THE HEBREW BIBLE AS LITERATURE Tod Linafelt
HEGEL Peter Singer
HEIDEGGER Michael Inwood

THE HELLENISTIC AGE
Peter Thonemann
HEREDITY John Waller
HERMENEUTICS Jens Zimmermann
HERODOTUS Jennifer T. Roberts
HIEROGLYPHS Penelope Wilson
HINDUISM Kim Knott
HISTORY John H. Arnold
THE HISTORY OF ASTRONOMY
Michael Hoskin
THE HISTORY OF CHEMISTRY
William H. Brock
THE HISTORY OF CHILDHOOD
James Marten
THE HISTORY OF CINEMA
Geoffrey Nowell-Smith
THE HISTORY OF LIFE
Michael Benton
THE HISTORY OF MATHEMATICS
Jacqueline Stedall
THE HISTORY OF MEDICINE
William Bynum
THE HISTORY OF PHYSICS
J. L. Heilbron
THE HISTORY OF TIME
Leofranc Holford-Strevens
HIV AND AIDS Alan Whiteside
HOBBES Richard Tuck
HOLLYWOOD Peter Decherney
THE HOLY ROMAN EMPIRE
Joachim Whaley
HOME Michael Allen Fox
HOMER Barbara Graziosi
HORMONES Martin Luck
HUMAN ANATOMY
Leslie Klenerman
HUMAN EVOLUTION Bernard Wood
HUMAN RIGHTS Andrew Clapham
HUMANISM Stephen Law
HUME A. J. Ayer
HUMOUR Noël Carroll
THE ICE AGE Jamie Woodward
IDENTITY Florian Coulmas
IDEOLOGY Michael Freeden
THE IMMUNE SYSTEM
Paul Klenerman
INDIAN CINEMA
Ashish Rajadhyaksha
INDIAN PHILOSOPHY Sue Hamilton
THE INDUSTRIAL REVOLUTION
Robert C. Allen
INFECTIOUS DISEASE
Marta L. Wayne
and Benjamin M. Bolker
INFINITY Ian Stewart
INFORMATION Luciano Floridi
INNOVATION Mark Dodgson and
David Gann
INTELLECTUAL PROPERTY
Siva Vaidhyanathan
INTELLIGENCE Ian J. Deary
INTERNATIONAL LAW
Vaughan Lowe
INTERNATIONAL MIGRATION
Khalid Koser
INTERNATIONAL RELATIONS
Paul Wilkinson
INTERNATIONAL SECURITY
Christopher S. Browning
IRAN Ali M. Ansari
ISLAM Malise Ruthven
ISLAMIC HISTORY Adam Silverstein
ISOTOPES Rob Ellam
ITALIAN LITERATURE
Peter Hainsworth and David Robey
JESUS Richard Bauckham
JEWISH HISTORY David N. Myers
JOURNALISM Ian Hargreaves
JUDAISM Norman Solomon
JUNG Anthony Stevens
KABBALAH Joseph Dan
KAFKA Ritchie Robertson
KANT Roger Scruton
KEYNES Robert Skidelsky
KIERKEGAARD Patrick Gardiner
KNOWLEDGE Jennifer Nagel
THE KORAN Michael Cook
KOREA Michael J. Seth
LAKES Warwick F. Vincent
LANDSCAPE ARCHITECTURE
Ian H. Thompson
LANDSCAPES AND
GEOMORPHOLOGY
Andrew Goudie and Heather Viles
LANGUAGES Stephen R. Anderson
LATE ANTIQUITY Gillian Clark
LAW Raymond Wacks
THE LAWS OF THERMODYNAMICS
Peter Atkins
LEADERSHIP Keith Grint
LEARNING Mark Haselgrove
LEIBNIZ Maria Rosa Antognazza

LEO TOLSTOY Liza Knapp
LIBERALISM Michael Freeden
LIGHT Ian Walmsley
LINCOLN Allen C. Guelzo
LINGUISTICS Peter Matthews
LITERARY THEORY Jonathan Culler
LOCKE John Dunn
LOGIC Graham Priest
LOVE Ronald de Sousa
MACHIAVELLI Quentin Skinner
MADNESS Andrew Scull
MAGIC Owen Davies
MAGNA CARTA Nicholas Vincent
MAGNETISM Stephen Blundell
MALTHUS Donald Winch
MAMMALS T. S. Kemp
MANAGEMENT John Hendry
MAO Delia Davin
MARINE BIOLOGY Philip V. Mladenov
THE MARQUIS DE SADE John Phillips
MARTIN LUTHER Scott H. Hendrix
MARTYRDOM Jolyon Mitchell
MARX Peter Singer
MATERIALS Christopher Hall
MATHEMATICAL FINANCE Mark H. A. Davis
MATHEMATICS Timothy Gowers
MATTER Geoff Cottrell
THE MEANING OF LIFE Terry Eagleton
MEASUREMENT David Hand
MEDICAL ETHICS Michael Dunn and Tony Hope
MEDICAL LAW Charles Foster
MEDIEVAL BRITAIN John Gillingham and Ralph A. Griffiths
MEDIEVAL LITERATURE Elaine Treharne
MEDIEVAL PHILOSOPHY John Marenbon
MEMORY Jonathan K. Foster
METAPHYSICS Stephen Mumford
METHODISM William J. Abraham
THE MEXICAN REVOLUTION Alan Knight
MICHAEL FARADAY Frank A. J. L. James
MICROBIOLOGY Nicholas P. Money
MICROECONOMICS Avinash Dixit
MICROSCOPY Terence Allen
THE MIDDLE AGES Miri Rubin
MILITARY JUSTICE Eugene R. Fidell
MILITARY STRATEGY Antulio J. Echevarria II
MINERALS David Vaughan
MIRACLES Yujin Nagasawa
MODERN ARCHITECTURE Adam Sharr
MODERN ART David Cottington
MODERN CHINA Rana Mitter
MODERN DRAMA Kirsten E. Shepherd-Barr
MODERN FRANCE Vanessa R. Schwartz
MODERN INDIA Craig Jeffrey
MODERN IRELAND Senia Pašeta
MODERN ITALY Anna Cento Bull
MODERN JAPAN Christopher Goto-Jones
MODERN LATIN AMERICAN LITERATURE Roberto González Echevarría
MODERN WAR Richard English
MODERNISM Christopher Butler
MOLECULAR BIOLOGY Aysha Divan and Janice A. Royds
MOLECULES Philip Ball
MONASTICISM Stephen J. Davis
THE MONGOLS Morris Rossabi
MOONS David A. Rothery
MORMONISM Richard Lyman Bushman
MOUNTAINS Martin F. Price
MUHAMMAD Jonathan A. C. Brown
MULTICULTURALISM Ali Rattansi
MULTILINGUALISM John C. Maher
MUSIC Nicholas Cook
MYTH Robert A. Segal
NAPOLEON David Bell
THE NAPOLEONIC WARS Mike Rapport
NATIONALISM Steven Grosby
NATIVE AMERICAN LITERATURE Sean Teuton
NAVIGATION Jim Bennett
NAZI GERMANY Jane Caplan
NELSON MANDELA Elleke Boehmer
NEOLIBERALISM Manfred Steger and Ravi Roy
NETWORKS Guido Caldarelli and Michele Catanzaro

THE NEW TESTAMENT
Luke Timothy Johnson
THE NEW TESTAMENT AS LITERATURE Kyle Keefer
NEWTON Robert Iliffe
NIELS BOHR J. L. Heilbron
NIETZSCHE Michael Tanner
NINETEENTH-CENTURY BRITAIN
Christopher Harvie and H. C. G. Matthew
THE NORMAN CONQUEST
George Garnett
NORTH AMERICAN INDIANS
Theda Perdue and Michael D. Green
NORTHERN IRELAND
Marc Mulholland
NOTHING Frank Close
NUCLEAR PHYSICS Frank Close
NUCLEAR POWER Maxwell Irvine
NUCLEAR WEAPONS
Joseph M. Siracusa
NUMBER THEORY Robin Wilson
NUMBERS Peter M. Higgins
NUTRITION David A. Bender
OBJECTIVITY Stephen Gaukroger
OCEANS Dorrik Stow
THE OLD TESTAMENT
Michael D. Coogan
THE ORCHESTRA D. Kern Holoman
ORGANIC CHEMISTRY
Graham Patrick
ORGANIZATIONS Mary Jo Hatch
ORGANIZED CRIME
Georgios A. Antonopoulos and Georgios Papanicolaou
ORTHODOX CHRISTIANITY
A. Edward Siecienski
PAGANISM Owen Davies
PAIN Rob Boddice
THE PALESTINIAN-ISRAELI CONFLICT Martin Bunton
PANDEMICS Christian W. McMillen
PARTICLE PHYSICS Frank Close
PAUL E. P. Sanders
PEACE Oliver P. Richmond
PENTECOSTALISM William K. Kay
PERCEPTION Brian Rogers
THE PERIODIC TABLE
Eric R. Scerri
PHILOSOPHY Edward Craig
PHILOSOPHY IN THE ISLAMIC WORLD Peter Adamson
PHILOSOPHY OF BIOLOGY
Samir Okasha
PHILOSOPHY OF LAW
Raymond Wacks
PHILOSOPHY OF SCIENCE
Samir Okasha
PHILOSOPHY OF RELIGION
Tim Bayne
PHOTOGRAPHY Steve Edwards
PHYSICAL CHEMISTRY Peter Atkins
PHYSICS Sidney Perkowitz
PILGRIMAGE Ian Reader
PLAGUE Paul Slack
PLANETS David A. Rothery
PLANTS Timothy Walker
PLATE TECTONICS Peter Molnar
PLATO Julia Annas
POETRY Bernard O'Donoghue
POLITICAL PHILOSOPHY David Miller
POLITICS Kenneth Minogue
POPULISM Cas Mudde and Cristóbal Rovira Kaltwasser
POSTCOLONIALISM Robert Young
POSTMODERNISM Christopher Butler
POSTSTRUCTURALISM
Catherine Belsey
POVERTY Philip N. Jefferson
PREHISTORY Chris Gosden
PRESOCRATIC PHILOSOPHY
Catherine Osborne
PRIVACY Raymond Wacks
PROBABILITY John Haigh
PROGRESSIVISM Walter Nugent
PROHIBITION W. J. Rorabaugh
PROJECTS Andrew Davies
PROTESTANTISM Mark A. Noll
PSYCHIATRY Tom Burns
PSYCHOANALYSIS Daniel Pick
PSYCHOLOGY Gillian Butler and Freda McManus
PSYCHOLOGY OF MUSIC
Elizabeth Hellmuth Margulis
PSYCHOPATHY Essi Viding
PSYCHOTHERAPY Tom Burns and Eva Burns-Lundgren
PUBLIC ADMINISTRATION
Stella Z. Theodoulou and Ravi K. Roy
PUBLIC HEALTH Virginia Berridge

PURITANISM Francis J. Bremer
THE QUAKERS Pink Dandelion
QUANTUM THEORY
John Polkinghorne
RACISM Ali Rattansi
RADIOACTIVITY Claudio Tuniz
RASTAFARI Ennis B. Edmonds
READING Belinda Jack
THE REAGAN REVOLUTION Gil Troy
REALITY Jan Westerhoff
RECONSTRUCTION Allen. C. Guelzo
THE REFORMATION Peter Marshall
RELATIVITY Russell Stannard
RELIGION IN AMERICA Timothy Beal
THE RENAISSANCE Jerry Brotton
RENAISSANCE ART
Geraldine A. Johnson
RENEWABLE ENERGY Nick Jelley
REPTILES T. S. Kemp
REVOLUTIONS Jack A. Goldstone
RHETORIC Richard Toye
RISK Baruch Fischhoff and John Kadvany
RITUAL Barry Stephenson
RIVERS Nick Middleton
ROBOTICS Alan Winfield
ROCKS Jan Zalasiewicz
ROMAN BRITAIN Peter Salway
THE ROMAN EMPIRE
Christopher Kelly
THE ROMAN REPUBLIC
David M. Gwynn
ROMANTICISM Michael Ferber
ROUSSEAU Robert Wokler
RUSSELL A. C. Grayling
RUSSIAN HISTORY Geoffrey Hosking
RUSSIAN LITERATURE Catriona Kelly
THE RUSSIAN REVOLUTION
S. A. Smith
SAINTS Simon Yarrow
SAVANNAS Peter A. Furley
SCEPTICISM Duncan Pritchard
SCHIZOPHRENIA Chris Frith and
Eve Johnstone
SCHOPENHAUER
Christopher Janaway
SCIENCE AND RELIGION
Thomas Dixon
SCIENCE FICTION David Seed
THE SCIENTIFIC REVOLUTION
Lawrence M. Principe
SCOTLAND Rab Houston
SECULARISM Andrew Copson
SEXUAL SELECTION Marlene Zuk and
Leigh W. Simmons
SEXUALITY Véronique Mottier
SHAKESPEARE'S COMEDIES
Bart van Es
SHAKESPEARE'S SONNETS AND
POEMS Jonathan F. S. Post
SHAKESPEARE'S TRAGEDIES
Stanley Wells
SIKHISM Eleanor Nesbitt
THE SILK ROAD James A. Millward
SLANG Jonathon Green
SLEEP Steven W. Lockley and
Russell G. Foster
SOCIAL AND CULTURAL
ANTHROPOLOGY
John Monaghan and Peter Just
SOCIAL PSYCHOLOGY Richard J. Crisp
SOCIAL WORK Sally Holland and
Jonathan Scourfield
SOCIALISM Michael Newman
SOCIOLINGUISTICS John Edwards
SOCIOLOGY Steve Bruce
SOCRATES C. C. W. Taylor
SOUND Mike Goldsmith
SOUTHEAST ASIA James R. Rush
THE SOVIET UNION Stephen Lovell
THE SPANISH CIVIL WAR
Helen Graham
SPANISH LITERATURE Jo Labanyi
SPINOZA Roger Scruton
SPIRITUALITY Philip Sheldrake
SPORT Mike Cronin
STARS Andrew King
STATISTICS David J. Hand
STEM CELLS Jonathan Slack
STOICISM Brad Inwood
STRUCTURAL ENGINEERING
David Blockley
STUART BRITAIN John Morrill
THE SUN Philip Judge
SUPERCONDUCTIVITY
Stephen Blundell
SUPERSTITION Stuart Vyse
SYMMETRY Ian Stewart
SYNAESTHESIA Julia Simner
SYNTHETIC BIOLOGY Jamie A. Davies
SYSTEMS BIOLOGY Eberhard O. Voit

TAXATION Stephen Smith
TEETH Peter S. Ungar
TELESCOPES Geoff Cottrell
TERRORISM Charles Townshend
THEATRE Marvin Carlson
THEOLOGY David F. Ford
THINKING AND REASONING Jonathan St B. T. Evans
THOMAS AQUINAS Fergus Kerr
THOUGHT Tim Bayne
TIBETAN BUDDHISM Matthew T. Kapstein
TIDES David George Bowers and Emyr Martyn Roberts
TOCQUEVILLE Harvey C. Mansfield
TOPOLOGY Richard Earl
TRAGEDY Adrian Poole
TRANSLATION Matthew Reynolds
THE TREATY OF VERSAILLES Michael S. Neiberg
TRIGONOMETRY Glen Van Brummelen
THE TROJAN WAR Eric H. Cline
TRUST Katherine Hawley
THE TUDORS John Guy
TWENTIETH-CENTURY BRITAIN Kenneth O. Morgan
TYPOGRAPHY Paul Luna
THE UNITED NATIONS Jussi M. Hanhimäki
UNIVERSITIES AND COLLEGES David Palfreyman and Paul Temple
THE U.S. CONGRESS Donald A. Ritchie
THE U.S. CONSTITUTION David J. Bodenhamer
THE U.S. SUPREME COURT Linda Greenhouse
UTILITARIANISM Katarzyna de Lazari-Radek and Peter Singer
UTOPIANISM Lyman Tower Sargent
VETERINARY SCIENCE James Yeates
THE VIKINGS Julian D. Richards
VIRUSES Dorothy H. Crawford
VOLTAIRE Nicholas Cronk
WAR AND TECHNOLOGY Alex Roland
WATER John Finney
WAVES Mike Goldsmith
WEATHER Storm Dunlop
THE WELFARE STATE David Garland
WILLIAM SHAKESPEARE Stanley Wells
WITCHCRAFT Malcolm Gaskill
WITTGENSTEIN A. C. Grayling
WORK Stephen Fineman
WORLD MUSIC Philip Bohlman
THE WORLD TRADE ORGANIZATION Amrita Narlikar
WORLD WAR II Gerhard L. Weinberg
WRITING AND SCRIPT Andrew Robinson
ZIONISM Michael Stanislawski

Available soon:

FIRE Andrew C. Scott
ECOLOGY Jaboury Ghazoul
MODERN BRAZIL Anthony W. Pereira
BIOGEOGRAPHY Mark V. Lomolino
ÉMILE ZOLA Brian Nelson

For more information visit our website
www.oup.com/vsi/

Philip Judge

THE SUN

A Very Short Introduction

OXFORD
UNIVERSITY PRESS

Great Clarendon Street, Oxford, OX2 6DP,
United Kingdom

Oxford University Press is a department of the University of Oxford.
It furthers the University's objective of excellence in research, scholarship,
and education by publishing worldwide. Oxford is a registered trade mark of
Oxford University Press in the UK and in certain other countries

First edition published in 2020

Impression: 2

Published in the United States of America by Oxford University Press
198 Madison Avenue, New York, NY 10016, United States of America

British Library Cataloguing in Publication Data
Data available

Library of Congress Control Number: 2020934353

ISBN 978-0-19-883269-0

Printed in Great Britain by
Ashford Colour Press Ltd, Gosport, Hampshire

Contents

Acknowledgements

I am thankful to all the scientists who have helped me to appreciate the workings of the Sun and stars. I want to thank especially Carole Jordan of Oxford University, whose generosity, wisdom, and example I have tried to follow over my career. I thank those of the High Altitude Observatory of the National Center for Atmospheric Research who have, one way or another, mentored me and allowed me to develop over three decades. I am particularly grateful to Dr Boon Chye Low for his continuing education of the remarkable world of magnetized plasmas. I thank my wife Terri Resley for her love, support, and encouragement. Without her, my sister Claire Judge, Dr Robert Rifkin, and a lot of luck, I would not have been here after 2003, and certainly not after 2004!

List of illustrations

Chapter 1
The Sun, our star

Our familiar Sun

It is no surprise that our ancestors worshipped the Sun. To the ancients, the Sun was the dependable source of light, heat, raw energy, and life itself. For civilizations everywhere throughout time, the rising of the Sun was, as a matter of daily experience, something to be wholly relied upon. Time itself was marked by the Sun's motions. No aspect of life was unaffected by the Sun. The Sun's steady brightness and regular motion marked seasons for planting, harvesting, lambing, and calving. Civilizations worldwide honoured the Sun with monuments that were built to endure. The ancients experienced and documented solar eclipses. Such moments caused uncertainty and consternation, leading to awe, fear, and wonder. The temporary darkness during total eclipses served perhaps to remind humanity that the heavens could not entirely be taken for granted. In some cultures, eclipses took on deep significance with important political consequences. In the Chinese 'Bamboo Annals' dating from *c.*300 BCE, an eclipse of 1059 BCE is recorded as having triggered actions leading to the fall of the Shang and rise of the Zhou dynasties.

Yet, in the hustle and bustle of our modern lives, like a devoted spouse, the Sun is all too readily taken for granted. A stockbroker might promise that a share's price will increase 'as sure as the Sun will rise tomorrow'. We have become accustomed to our star.

Perhaps our nonchalant acceptance arises from our modern understanding of the mechanics of the solar system, developed first by Isaac Newton (1642–1727), who had synthesized theories of motion and gravity based upon astronomical observations and calculations of Nicolas Copernicus (1473–1543), Tycho Brahe (1546–1601), Johannes Kepler (1571–1630), and Galileo Galilei (1564–1642). We know that the Sun will rise tomorrow owing to Earth's rotation, a consequence of Newton's laws of motion. Our 20th-century knowledge of nuclear processes and energy transport gives us faith that it will also shine. We can easily forget that the Sun is the source of almost all energy that we can readily use, from fossil fuels, winds and waves, to solar panels.

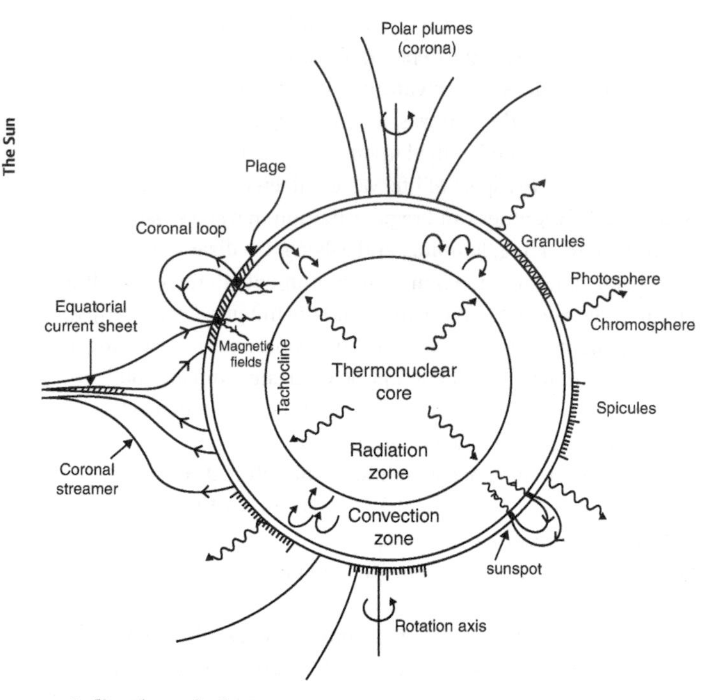

1. A slice through the Sun.

So long as the workings of the Universe are the same tomorrow as they are today, we can live our lives simply knowing that the Sun will be our faithful companion. This, a modern 'act of faith', is based upon the same experiences as the ancients. For, if this had not been the case over the almost unthinkably long time of several aeons, the large and intricate forms of life on Earth, such as you and I, could not have come into being.

Unlike some acts of faith, modern understanding of nature (an endeavour called 'science') is supported by a vast collection of evidence, and centuries of rejection of unsatisfactory pictures, weeded out by experience and by pitting ideas against observations. The result of such endeavours in understanding the Sun is summarized in Figure 1, which will serve as a reference for the rest of this chapter.

This is a book about the science, the workings of the Sun, about solar physics.

The importance of the Sun

Why is the Sun important? It is, after all, a nondescript star, differing little from billions in the Milky Way galaxy. It is a star of typical mass, typical colour and luminosity, a middle-aged star. It is of spectral type G in the range O, B, A, F, G, K, and M that is a sequence of decreasing mass and temperature. Yet the Sun *is* special and of interest, simply because of our proximity to it. From Earth's orbit, the Sun is over 270,000 times closer than α Centauri A, the nearest night-time star that is otherwise Sun-like. Our closeness allows us to observe, measure, and document its behaviour in detail.

Table 1 lists some physical properties of the Sun as well as its classification among the stars. The notation 'G2' in the spectral type refers to the Sun's colour and approximate temperature in the

Table 1. Some basic properties of the Sun

Mass ($M_\odot$)	2×10^{30} kg
Radius ($R_\odot$)	700,000 km
Distance	150,000,000 km $\equiv$ 1 AU
Luminosity ($L_\odot$)	4×10^{26} W
Irradiance at Earth	1.365 kW/m^{-2}
Average rotation period	27 days
Stellar spectral type	G2 V

Notes: kg=kilograms, AU=astronomical unit, km=kilometres, W=Watts, kW=kiloWatts, kW m^{-2}=kiloWatts per metre^2.

stellar O through M sequence, running for G-type stars from G0 to G8. The 'V' denotes the size of the star, or its 'luminosity' class among the stars. It classifies the Sun as a normal dwarf star, currently on the 'main sequence' (Chapter 2).

The irradiance at Earth is the amount of radiative power received across 1 metre square (m^2) at Earth's orbit, over 1 kiloWatt per m^2 (kW/m^2). Not only does this make the Earth habitable with a comfortable average temperature of 14 Celsius, but if we could convert 100 per cent of the radiative power of the Sun collected by an area of just 0.4 m^2 into electrical power, it would satisfy the current energy needs of an average household in the UK continuously.

The Sun's proximity also makes it important as a laboratory for studying basic physical phenomena that can occur under conditions totally unreachable in our laboratories. Solar studies inform us of nature operating on the enormous scales encountered across the Universe. It exhibits remarkable phenomena that we take for granted, such as sunspots, the corona, flares, the solar wind, and coronal mass ejections (CMEs) to be discussed below. Similar phenomena occur in different guises in many astrophysical plasmas in which plasma motions and magnetic fields interact. Yet we could not anticipate any of these phenomena from first principles. The Sun is thus a proving ground where we can test theories of 'magnetized plasmas' of importance also in our quest to find safe power from nuclear fusion. Using the Sun as our laboratory, we can try new measurement techniques in astronomy, and pit them against theories and models, in our quest to understand the entire Universe.

The Sun and stars are in a different state of matter than the solids, liquids, and gases with which we are familiar. They consist of ionized particles that are in the *plasma* state. Just as it takes energy to break chemical bonds to make a liquid from a solid (melting), and a gas from a liquid (evaporation), the addition of considerably more energy separates the electrons from their atomic nuclei, to form an electron-ion plasma. Plasmas are the natural state of matter when the energy per particle—proportional to their temperature—is high enough to keep electrons separated from their parent atoms. Most objects in astronomy bright enough to be seen are in the plasma state. Along with charged particles, plasmas are flooded with light that is emitted and absorbed by the moving charges. We can consider a plasma to consist of electrons, ions, and photons, the latter being small quanta of light. As we will see, the large energy densities required to make these astronomical plasmas arise naturally in astronomy from the inexorable effects of gravity, after matter condensed out of the raw energy of the Big Bang.

The defining property of plasma is that it is made up of free charges—electrons and ions. Electrical charges were familiar to philosophers from 600 BCE. But its effects were first experimentally quantified far later, by Charles-Augustin de Coulomb (1736–1806). He developed the concept of an 'electric field', in which each electrical charge endows the space around it with a force that acts on other charges. The lines of force begin and end at electrical charges.

The freedom of charged particles in a plasma to move between random collisions means that the plasma is an excellent *electrical conductor*. Just as the electrons easily move along household copper wires to form a current, an orderly relative motion of free electrons and ions in a plasma constitutes an electric current inside the plasma itself. The Sun's corona conducts electricity like copper, but plasma is more like a fluid than a wire. Mercury is familiar to us as a liquid on Earth's surface, with a conductive efficiency of about one-fiftieth of copper, and indeed laboratory experiments with liquid metals have helped inform us about the workings of the Sun. In contrast, seawater supports electrical currents carried by electrons removed from and attached to ions of sodium and chlorine in solution. These heavier ions are far less mobile than electrons, and so seawater has about one-ten-millionth of the conductive efficiency of copper.

Both magnetic and electric fields exist in the Earth's atmosphere, allowing birds to navigate and sparks to occur through lightning. But in the Sun, the electrons are so mobile that any large-scale *electric* fields are almost immediately shorted out, just as in a copper wire. Any *magnetic* fields survive intact against short-outs, because no-one has found evidence for any magnetic charges, called 'magnetic monopoles'. Based upon historic experiments by Coulomb, Jean-Baptiste Biot (1774–1862), Felix Savart (1791–1841), André-Marie Ampère (1775–1836) and Michael Faraday (1791–1867), James Clerk Maxwell (1831–79) 'unified' electricity with magnetism. Experiments show that magnetic fields are generated not by monopoles, but by electric currents, that is, the relative motion of positive and negative charges. In turn, magnetic fields exert forces only when charges and field are in relative motion.

In the absence of monopoles, magnetic field lines have no beginning and no end. Only two kinds of magnetic field lines can satisfy this condition. One is that the magnetic field lines make up complete loops, such as those threading through and around bar magnets (see Figure 7). The other kind is in an 'ergodic' state where field lines go on forever throughout the universe. The former concept is most useful for discussing a finite object like the Sun.

Electric and magnetic charges

While electric fields begin and end at electric charges, magnetic fields circulate around electric currents. Maxwell's theory is entirely symmetric with regard to electric and magnetic fields, with just one exception: it contains no magnetic charges. So, magnetic lines of force have no beginning, no end. If magnetic charges were abundant, they too would in principle move, just like the electrons, to short-out their fields. But this is not the case. In an elegant book *Conversations on magnetic and electric fields in the cosmos*, Eugene Parker (1927–) concludes that abundant magnetic charges ('monopoles') would short-out any magnetic fields within plasmas. But in their apparent absence in our Universe, we see that *a body of plasma like the Sun can sustain large-scale magnetic fields but not* ***large-scale*** *electric fields.*

The Sun is important also because it is not quite as benign as some might believe. The solar corona emits a steady stream of ultraviolet (UV) and X-radiation, as well as plasma emerging as the solar wind. The interactions between dense solar plasma beneath the visible surface and the magnetic fields that generate the corona can eject intense and unpredictable bursts of high energy radiation, jets, plasma, and magnetic fields into interplanetary space. The Sun is a machine that converts a small but important fraction of its benign power into variable energetic radiation, magnetism, and particles. It is fair to say that today *the biggest problems in solar physics concern the dynamical interactions between solar plasma and its magnetic fields.*

In passing the Earth, solar outbursts disturb Earth's protective magnetic field, causing problems for electrical infrastructure, though leading to beautiful aurorae. Our increasing technology-dependence makes our way of life vulnerable to sustaining damage as a result of the poorly-understood workings of our otherwise friendly neighbour.

There are plenty of reasons, then, why the Sun is important and worthy of further study.

Sunlight and the beginnings of astrophysics

The publication in 1687 of *Principia* by Sir Isaac Newton is as good a place as any to mark the beginning of physics—the subject concerned with the quantitative understanding of elementary natural processes. But it was not until the 19th century that the seeds of *astro*-physics—concerned with the nature of remote heavenly objects, as opposed to just their motions—were planted. In 1814, Joseph von Fraunhofer (1787–1826), following work with prisms by Newton and by William Hyde Wollaston (1766–1828) in 1802, re-examined the solar spectrum. The solar light entered Fraunhofer's spectroscope through a slit, using lenses and a prism to enhance the splitting of light into constituent colours. In this way he was able to show that the spectrum of the Sun consisted of a bright background with a large number of darker dips at particular wavelengths.

In 1821, Fraunhofer developed a new spectroscope based on *diffraction*, to achieve a stronger splitting of light, dispersing it into more than just broad constituent colours. The trick took advantage of the wave nature of light. Fraunhofer's 'dips' (now called spectral lines) were labelled alphabetically ranging from red to violet. These lines have been commemorated on a stamp, as shown in Figure 2. The smoothly changing background in the stamp shows the 'continuous' thermal solar spectrum, going from infrared (IR) on the left, to UV on the right. Figure 3 shows the spectral brightness as a function of wavelength as a line plot, with colours marked, showing only the continuum.

The dips in the spectrum are the dark vertical lines in the lower half of the stamp. They are a persistent solar feature. Their origin remained a puzzle, until breakthrough laboratory experiments were made by Robert Bunsen (1811–99) and Gustav Kirchoff

2. A stamp issued in 1987 to commemorate the work of Joseph von Fraunhofer, showing some of the dips in the solar spectrum. These are named 'spectral lines' although the word line only originates from the geometry of a linearly-shaped slit used at the entrance to a spectroscope.

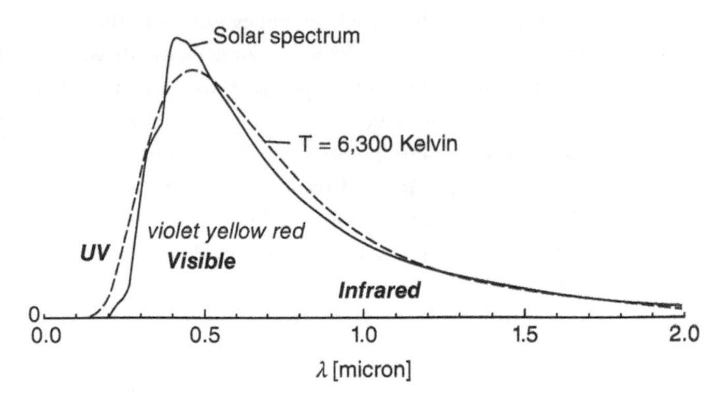

3. The brightness of the Sun is plotted against wavelength of light in microns (1 micron is a millionth of one metre). UV light is at the left, IR at the right, the opposite sense to Figure 2. The dashed curve shows a 'black body' at a temperature of 6,300 Kelvin. Almost all the solar radiation emerges between 0.2 and 2 microns, spanning the 0.4 to 0.8 micron region visible to our eyes.

(1824–87). In 1860 they demonstrated that the solar dips coincided with wavelengths of bright emission seen in some flames. For example, two solar lines in the yellow region (labelled D1 and D2) appeared when sodium was introduced into the flames. The coincidence in wavelengths hinted that sodium was present on the Sun. John Dalton (1766–1844) had found that matter is assembled of 'elements', and Bunsen and Kirchoff had demonstrated that compounds of such elements, seen under conditions of combustion in gases, had these tell-tale spectral fingerprints. Soon after solar spectra were scrutinized, bright star spectra were also compared, revealing diverse sets of spectral lines. Thus began the era of modern astrophysics, the *era in which we could begin to quantify the nature of astronomical bodies* remote from our laboratories.

For astronomers, spectral lines were the 'DNA' of their time. Just as DNA was to set genetics on a firm and quantitative basis a century later, spectroscopy allowed astronomers to build a firm foundation for examining the structure and evolution of the Universe. In fact, on occasion this 'DNA' of some elements was seen first not in the laboratory but in spectra of the Sun and other astronomical objects. A third solar line (D3) was speculatively associated with a new element 'helium' after it was seen in emission at a wavelength different from D1 and D2 in 1868, by Norman Lockyer (1836–1920) and Edward Frankland (1825–99). The intrepid Pierre Janssen (1824–1907) reported the emission line earlier, during and outside of a total eclipse, but had assumed it to be sodium. It was only later that it was identified in the laboratory by chemists Per Teodor Cleve (1840–1905) and Nils Abraham Langlet (1868–1936), while studying the uranium ore cleveite. Yet it is the second most abundant element in the Universe! At the eclipse of 1869, Charles Augustus Young (1834–1908; he died on the day of a total solar eclipse) and Civil War surgeon, astronomer William Harkness (1837–1903) independently observed visually the 'green coronal line'. The line's origin was unknown and some believed it to belong to a new element, 'coronium'. There followed a period of seventy years of debate until it was finally identified by Bengt Edlén (1906–93), along with several other lines of highly ionized complex atoms.

Fluids, gases, plasmas, and magneto-hydrodynamics

Motions in our atmosphere are described using fluid mechanics. Solar physics uses the same description. In both cases additional processes are added as needed (e.g. cloud condensation in our atmosphere), but large-scale motions are best treated using the so-called fluid approximation.

In physics, a fluid describes well-defined averages of numbers of particles over finite volumes. We speak of the number of atoms per cubic metre, and their average speeds. Temperatures measure the spread of speeds of individual atoms around the local averages. Equations are solved for these averages, continuous quantities first described mathematically by Leonhard Euler (1707–83). All physical processes occurring inside the small volumes (like collisions between particles) are not addressed in the fluid approach, instead other methods such as kinetic theory are brought to bear to define necessary large-scale properties such as pressure, viscosity, and conductivities (like heat or electrical conduction) inside the fluid. Even in this case, classical mechanics describes the small-scale motions and interactions well in almost all conditions, except during the latest stages of evolution of the Sun, when quantum effects become critical.

Euler's work builds on experiments and theory extending back to Archimedes. The price paid by Euler's averaging is that the formulation becomes *non-linear*, leading to difficult problems such as turbulence. But this is more than worth it, for otherwise we would have to trace trajectories of an impossibly large number of interacting particles.

The simplest description of solar plasma is as a *magnetized fluid*. This fluid can sustain large-scale magnetic fields, but not electric fields. The subject is called *magneto-hydrodynamics* or MHD.

Over the next century, advancements in solar physics came hand-in-hand with several revolutionary advances in physics, including statistical mechanics (Ludwig Boltzmann, 1844–1906; Maxwell), the unification of electricity and magnetism by Maxwell, the birth and evolution of quantum mechanics, and Albert Einstein's (1879–1955) theories of relativity. In the early 1920s, Meghnad Saha (1893–1956) developed an extension to the Maxwell-Boltzmann statistical theory to include the effects of quantization on the 'ionization state' of a gas, or plasma. The state of ionization describes how many electrons are bound to a given atom. Carbon (C), for example, normally has six electrons in orbit around the nucleus which itself has six protons and (usually) six neutrons. If we remove one electron, we have C^{+}, written because we have only five negative electrons surrounding the six proton charges in the nucleus, leaving a net positive charge of one proton. Removing another electron leads to a state of carbon labelled C^{2+}, and so on. Each removal of an electron requires energy; temperature measures the amount of energy in the motions of each particle in a plasma, and Saha was able to derive a new equation describing how, at each temperature and density, the ionization stages are determined.

For astronomy, Saha's development proved crucial because the statistical theory of plasmas was, until then, incomplete, there being no theory that included the degrees of freedom associated with free electrons. Into the revolutionary atmosphere of physical advances of the 1920s came a remarkable student, English-born Cecilia Payne-Gaposhkin (1900–79), supervised by the director of Harvard College Observatory, Harlow Shapley (1885–1972). In her 1925 PhD thesis, armed with Saha's statistical theory, Payne-Gaposhkin examined the amount of the star's light that is absorbed in the spectral lines. She found that it depended critically on the *state of ionization*, not just the abundance of the elements. Her analysis led her to conclude that the Sun was composed mostly of hydrogen. This was such a shocking result at the time that eminent physicist Henry Norris Russell (1877–1957) advised that she not adhere to her findings. Initially she described her conclusions as spurious. But her work was proven right, including by Russell himself a few years later. By the mid-1920s, the subject of astrophysics was set to explode. Our modern theory of stellar evolution, based upon Payne-Gaposhkin's work in 1925, and with the Sun at stage centre, is surely one of humanity's great intellectual accomplishments.

Solar structure

Figure 1 sketches the overall structure of the Sun. The story behind this picture is remarkable, encompassing and driving some of the giant leaps in physics made during the 20th century. The *core* of the Sun, consisting of the energy-generating centre and a region of radiative energy transport, is surrounded by the *convection zone*. While neither region is directly observable using light or other kinds of radiation, we believe that their basic structure is well-understood. Immediately above the convection zone is the visible 'surface' called the *photosphere*. It too is fairly well-understood, subject to some intricacies discussed later. The photosphere, named after the Greek word for light, $\phi\omega\zeta$ (*phos*), spans the region where essentially all of the visible light escapes from the Sun. Above this region lie the *chromosphere* and *corona*, two thermally stable regions between which there lies a more dynamic *transition region*. The latter three components are visible only during eclipse or using light of very particular wavelengths. Finally, the solar wind emerges from parts of the corona where the Sun's magnetic fields open into space.

The solar 'surface' is a diffuse region from which visible light emerges. It is neither a liquid nor a solid surface, instead it is akin to our atmosphere which gradually gets thinner with height. It appears sharp to our eyes because it has a thickness less than 300 kilometres (km), compared with the solar radius of 700,000

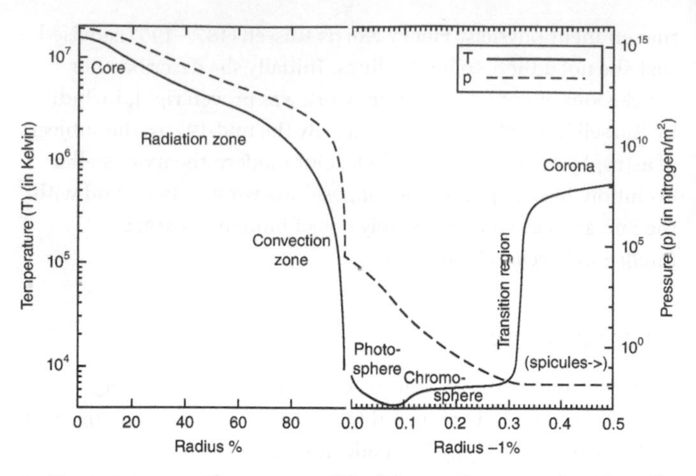

4. Temperatures and pressures of the Sun are shown as a function of distance from Sun centre. Temperatures are in Kelvin. To convert to Celsius simply subtract 273 from the Kelvin scale. Only those regions in the right half of the figure are directly observable.

km. The Sun's thermal structure, including this thin surface, is shown as a function of distance from the centre in Figure 4. While the observable atmosphere is geometrically 'thin' almost half of the pressure drop from Sun centre to interplanetary space takes place in observable regions.

The Sun's global internal structure is stable. It steadily loses energy to space that is generated by nuclear fusion in the core. Unlike our fusion reactors, the solar core is self-regulating, acting as a thermostat. Any local excess fluctuation of energy heats the particles and radiation, producing an enhanced pressure. Neighbouring regions are pushed around, and this pushing does work, sapping the energy from the overheated particles and cooling them. This is a classic negative-feedback loop. On average, the temperature is largest where the fusion is taking place. Common experience tells us that heat moves from hot to cold environments, always. In physics, this is expressed by the Second Law of Thermodynamics. Thus the temperature drops with distance from the centre.

The Sun's innate stability is not confined to the core. At Earth we measure changes in brightness of typically 0.04 per cent, related to the variable solar magnetic field. We don't know if they are compensated by brightness variations in directions out of Earth's orbital plane. But the supply of energy deep within the core indeed changes slowly, the 'main sequence lifetime', measuring the time during which core hydrogen is fused to form helium while maintaining a steady balance against the collapsing force of gravity, is ten billion (or 10^{10}) years.

A steady supply of energy does not mean that the output from the surface must also be steady, for if we consider a spherical shell inside the Sun (like a coconut shell), physics dictates that *on average* the energy entering the shell from below and leaving it from the top is precisely the same. If this were not the case, excesses or deficits of energy would build up in particular shells. Such irregularities can occur, but they quickly become smoothed out as the Sun relaxes to its steady state on small time scales. Differences in pressure are communicated inside a fluid by *sound waves* which, in the solar interior, take only a couple of hours to travel a solar radius.

Other than a possible small magnetic reservoir which in principle can store energy in internal magnetic fields, there are no thermal reservoirs in which to store and release energy later. But not all stars behave like the Sun. Deficits and excesses of energy can be spontaneously generated by certain instabilities. Cepheid variables, for example, vary greatly in brightness with a period correlated with peak brightness, giving astronomers a way to measure distance across the Universe from their observed periods and apparent brightnesses. The reservoir of energy in this case is in the latent energy of ionization of hydrogen and helium.

Why this structure?

In the solar core, the mean distances travelled by electrons and atomic nuclei are small before they collide with another particle. The particles are all charged and they all feel the electric effect of the charges in their immediate vicinity. These electrical interactions on tiny scales limit the net speed at which energy generated in the core can be transported by particle collisions. This process is none other than ordinary heat conduction, occurring inside a plasma. But the plasma is also bathed in radiant energy, carried by photons. Maxwell's theory of electromagnetism shows that photons are generated whenever a charged particle's velocity changes, such as occurs by the particle collisions in a plasma. That photons carry energy is evident because we feel the Sun's 'heat' on our skin. Inside the solar plasma, the distances travelled by photons are far larger, enough that they efficiently carry all the energy outwards, in what is called the 'radiative zone'. The inner 70 per cent of the Sun (by radius) transports energy like this.

But then something interesting happens as a result of the steady drop in temperature (see Figure 4), and in particular a change in the temperature gradient (the change in temperature with radius). At the lower temperatures, some of the atomic nuclei attract passing electrons, forming an ion with one or two electrons attached. These ions can be ionized by radiation, and some of the energy-carrying photons start to get *blocked* as a result. But, as shown by Einstein in 1905 in his Nobel-winning paper on the 'photo-electric effect', ionizations can only occur if light behaves as a particle, a photon, with a certain amount of energy. The effect necessitated abandoning classical mechanics in favour of quantum mechanics. The process, called 'photo-ionization', significantly reduces distances over which the photons can move. It also takes the energy from the photon and stores it in the plasma in two forms: the nucleus has one fewer electron than before and considerable energy is needed to separate the electron and nucleus; the electron carries energy in its motion. When enough photo-ionizations occur, radiation can no longer carry the needed energy to the surface from the ongoing fusion beneath. Locally, the plasma faces a dilemma: it cannot transport the energy through by radiation or conduction.

The fusion region inside the core knows nothing of this dilemma. It continues to release energy as normal. However, when this process begins, there is a build up of energy. One can imagine a narrow shell inside the Sun, like the coconut shell. The dilemma occurs because the radiation out of the outer part of the shell cannot balance the radiation coming in from below. The plasma heats at the edge nearer the core, but the energy cannot be carried upwards by radiation. The result is that the locally heated plasma must start to expand against the ambient pressure. It most easily expands upwards against the lower average pressures above. In the continuing pile-up of energy beneath, this process must continue, until the energy carried by upward motion can, on average, balance the incoming radiative/fusion energy. The transport of energy by such motions is called *thermal convection*. In the Sun, we therefore have a *convection zone* that happens to begin at 70 per cent of the solar radius (see Figure 4). The radius at which convection begins varies between different stars: it starts deeper in the core for cooler stars (of type K and M), and it disappears for hotter stars (O-, B-, and A-type stars).

Thermal convection is familiar as the unsteady, unpredictable updraughts that lead to thunderstorms. The air, warmed by the overheated ground below, is a very poor conductor of heat and electricity, and it is largely (but not completely) transparent to radiation. So the air must convect when ground heating exceeds a certain amount. These upward motions separate electrical charges in the air by friction (nothing other than the net effect of collisions between molecules and particles), in a slow build up of electrical energy that later can lead to sudden release as lightning. This does not happen in the Sun, because the Sun's plasma is electrically conducting. Sudden releases of energy do occur in *solar flares*, but through the slow build up of magnetic energy in the form of electrical currents, not by the gradual separation of electrical charges.

Convection is intrinsically *turbulent*. Usually this is not good news. Werner Heisenberg (1901–76) is reputed to have said, 'When I meet God, I am going to ask him two questions: Why quantum mechanics? And why turbulence? I really believe he will have an answer for the first' (this quotation has also been attributed to mathematician Horace Lamb; 1839–1934). Convection arises because warm fluid is lighter than cool fluid; it is driven by the inability of radiation to carry the solar flow of energy. It is a local instability, occurring on scales much smaller than the Sun. Therefore, the average effects of convection on the solar structure can be accurately modelled even though the individual cells are unpredictable, just like thunderstorm cells.

Eventually, the solar energy flow starts to become radiative again near the surface. This change is easier to appreciate. The distance travelled by the photons increases in response to the lower densities, and photons begin to flood freely into space. This change from convective to radiative energy transport defines the *solar photosphere*. The uppermost parts of the convection cells cool rapidly as the radiation escapes, forming a characteristic cellular surface structure called *granulation*, illustrated in Figure 6. The photosphere, the visible surface, is a diffuse layer about 300 km thick. All but at most 0.01 per cent of the power generated in the core is radiated away, the paltry remainder is used to heat and drive the overlying layers. The photosphere is *the* structure that radiates away essentially all the nuclear energy generated in the core. As such it has the right average temperature (6,000 Kelvin (K)) and forms at the right radius (700,000 km) to get rid of all this energy. Other stars have very different photospheric properties depending on their mass and evolutionary state, which fix the energy generation rate in the cores.

If the Sun radiated according to the laws of equilibrium thermodynamics, developed by Boltzmann, Maxwell, and Max Planck (1858–1937) between 160 and 110 years ago, it would produce a thermal spectrum (a 'black body' spectrum) which peaks at wavelengths of light visible to us. Figure 3 shows that, at visible wavelengths, the Sun actually radiates closely to a black body. Of course, our eyes' sensitivities to these brightest parts of the solar spectrum are no coincidence. We have evolved bathed in sunlight to be able to see as clearly as possible, and the Earth's atmosphere is almost transparent to visible wavelengths. The Sun therefore appears to us as an incandescent body with a temperature around 6,000 K.

The visible photosphere is now rather well-understood. Numerical computer models succeed in reproducing many salient observations of convection and, to some degree, magnetic fields permeating the surface. But the reader may rightly wonder about how we might validate, or refute, our models of the Sun's invisible interior. At least two more-or-less direct observations tell us that we are on the right track. First, certain kinds of neutrinos emerging directly from the solar core have been detected at the Earth. These ghostly particles, first postulated by Wolfgang Pauli (1900–58) in 1931, are released during nuclear decay. They interact very weakly with matter, and once generated, they emerge unattenuated from the solar core. Upon reaching the Earth, they can be detected, interacting just strongly enough for patient experimenters to accumulate data over years. After a decade or so of intense debate concerning a discrepancy between predictions of numbers of solar neutrinos and those measured at Earth (the so-called 'solar neutrino problem'), the solar models were vindicated. Second, although light waves cannot penetrate deeper than the solar photosphere, *sound* waves can. In response to kicks from random convective motions, the native elasticity of the Sun makes it vibrate throughout, like a jelly. These vibrations are observable as partly coherent patches of oscillations in brightness and velocity at the visible surface. Just as we are able to say something about the inside of, say, a bell, from the sound it makes, we can infer conditions that affect sound in the Sun's interior from measurements at the surface. We will meet these motions when we address 'helioseismology'. With such measurements we are able to map the Sun's inner temperature, fluid velocities, and some related quantities, as functions of radial distance, latitude, and sometimes longitude. Happily, these very different probes of the solar interior give us a consistent picture of the global structure, to a good approximation.

So, the structure shown in Figures 3 and 4 results from burning hydrogen in the core, and transporting energy outwards through the radiative zone, convection zone and photosphere. Heat flows from the hot interior to the cool surface, where the photosphere radiates freely into space. The emergent light peaks in the yellow region of the spectrum.

The solar structure above the photosphere is an area of active research. The properties shown in Figure 4 are ill-defined averages of complicated and dynamically-evolving structures, as we shall see.

The observable Sun

The observable regions of the Sun include the photosphere and the overlying features, broadly defined to include the chromosphere, corona, the intervening transition region, and the solar wind. They span an enormous range of pressures; Figure 4 shows that almost half of the pressure drop from Sun centre to interplanetary space takes place in the observable regions. The observed structures have historically been studied in great detail, often before we had an elementary idea of what they might be. As in other areas, this has led to a menagerie of jargon. Spots on the Sun were regularly observed beginning with Galileo and Jesuit Christoph Scheiner's (1576–1650) work three centuries before George Ellery Hale (1868–1938) uncovered their magnetic nature. Before Hale's time, sunspots were attributed to the passage of planets, solar mountains, or holes in luminous cloud decks. Even after Galileo's time, their solar origin was questioned by religious arguments concerning the purity of the heavens. Nowadays, as well as spots, we read of spot umbrae, their penumbrae, pores, granules, supergranules, faculae, plages, fibrils, filigree, spicules, jets, sprays, moustaches, bushes, rosettes, loops, prominences, filaments, and the list goes on. But such poetic words can cloud and confuse our physical understanding.

These complicated-looking phenomena arise for just *two physical reasons*. First, just below the visible surface, 100 per cent of the emerging energy is carried by the convective motions of the plasma, an electrically conducting fluid, and, second, the surface is permeated by magnetic fields emerging from beneath. The sub-surface convection churns up the fluid which, being a good conductor of electricity, twists and turns the magnetic fields around. In turn the fluid in the Sun begins to be forced by magnetic stresses in a back-reaction of the magnetic field, eventually working to change the fluid motions. The coupling between the fluid and the magnetic fields is a central theme in this book. It lies at the heart of what makes the Sun so intriguing to modern science and important to society.

Sunspots are just the most obvious manifestation of the effects of magnetism (Figure 5). They are comprised of large conglomerations of strong magnetic fields. Even the fearsome convective motions, carrying the energy from below, are suppressed by the intense magnetic forces near the surface, explaining why sunspots appear *dark*. Sunspots are the big bullies of the solar atmosphere and the convection just beneath. Their magnetic nature was first identified by Hale in 1908, using a

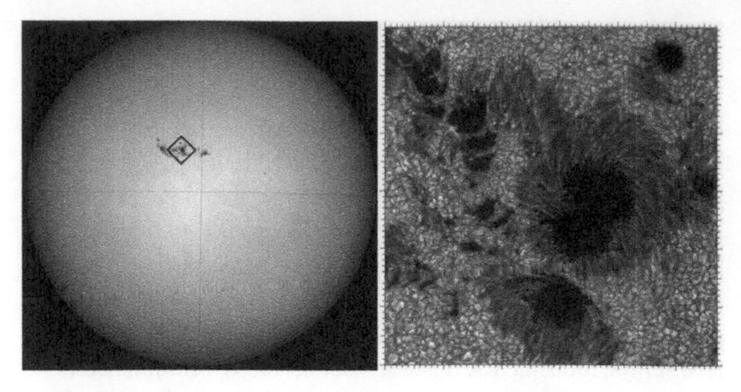

5. The Sun as it appears to the eye through a small and large telescope, highlighting a group of sunspots. The box on the left is the region shown on the right.

discovery by Dutchman Pieter Zeeman (1865–1943) of the magnetic splitting of spectral lines in the laboratory. In the very cores of sunspots, the magnetic fields are so intense that they resist and locally hinder convection, stopping some of the convective energy from reaching the surface. Sunspot cores, 'umbrae' (*shadows*; single: 'umbra') are therefore cooler and hence darker than the surrounding regions. Immediately surrounding the core, there are brighter regions called 'penumbrae' (*almost shadows*), regions dimmer than the ordinary surface but where striations trace out the outer magnetic fields of sunspots that are a complicated mix of horizontal and vertical magnetic field, and everything in-between.

The question of where the energy diverted by the dark sunspot goes is one of current interest. We might think that the radiation received at Earth would drop when the dark sunspots are present. But the answer is more complicated, because at the peripheries of the spots, outside the penumbrae, we find bright regions, called 'faculae' (*little torches*). Magnetism in spots is so intense that it exerts a pressure comparable to the plasma pressure of the fluid (see Figure 7 to visualize magnetic pressure). To keep in pressure balance with their environment, the plasma pressure inside a spot is considerably lower than that outside. Therefore less material is present in the umbrae to absorb and emit radiation, and so one sees *deeper* into the Sun in a spot. In 1769, Alexander Wilson (1714–86) noted that, when observed near the solar limb, that is, the very edge of the Sun's disc, sunspots appear to be *depressed* below the neighbouring photosphere. The effect, known as the 'Wilson depression', with a depth of a few hundred to 1,000 km, was unexplained for almost two centuries, when in 1965 Robert John Bray (1929–) and Ralph Edwin Loughhead (1929–) published the explanation given, in outline, above. The depression is not small—the depth often exceeds the entire thickness of the visible photosphere.

The Sun's magnetic fields outside of sunspots are weaker and more dispersed (Figure 6). These weaker magnetic fields exert smaller forces on the ions, electrons of the plasma, and so they are easily distorted by granular and other motions. The Sun's photosphere is covered with three million granules at any given time. They are the uppermost parts of convection cells (Figures 5 and 6). The hot, bright centres of granules move upwards, the cool, dark regions move downwards, carrying essentially the entire energy output of the Sun just below the visible surface. Individually the granule stuctures come and go in eight to twenty minutes. They are just the visible manifestation of the convection which transports the nuclear energy over the outer 30 per cent or so of the Sun (by radius) to this visible surface. As we shall soon see, any magnetic fields entrained in the granules must follow the fluid motions. The hot, upwelling fluid rapidly loses energy by radiation. It cools, ceases rising, and is forced sideways to the edges by more warm fluid emerging from beneath. At the cell edges the fluid rapidly descends beneath the surface. These motions drag weak magnetic fields towards the dark edges, where they become concentrated to form the narrow bright features that are observed (see Figure 6). Like the flotsam and jetsam that collects over drains, the magnetic fields are found at the down-flow regions between the up-welling granules. In Figure 6 the dark downflows are indicated,

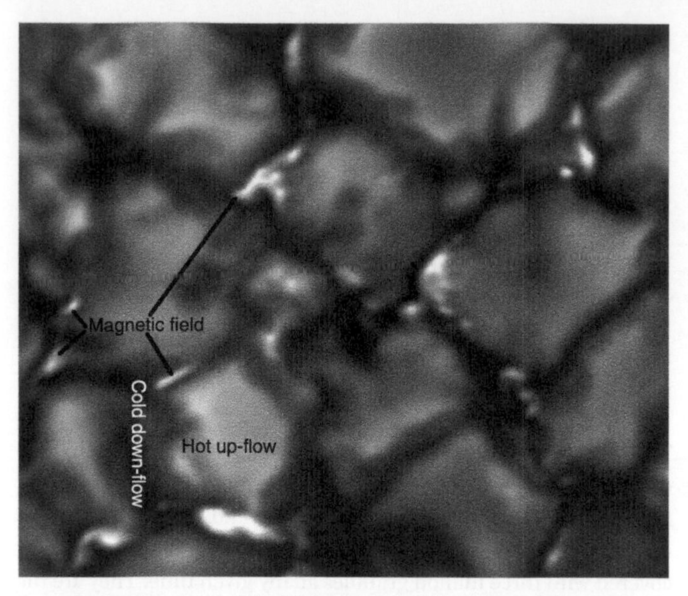

6. A small part of one of the highest resolution images of the Sun's photosphere is shown, covering an area of just thirty millionths of the solar disc. The large cells seen are granules. Ten complete granules are visible in the image. Their bright centres are where hot fluid has risen from beneath, carrying the bulk of outward flow of solar energy. The narrowest features are roughly 60 km across.

some of which contain brighter, narrow meandering structures where convoluted whirls and streams of magnetic fields are found. In contrast with sunspots, these magnetic fields make the dark lanes bright, because the plasma inside the magnetic fields is more transparent. But in contrast with the sunspots, convection is not suppressed in these small structures, and we can see deeper, where the plasma is hotter and brighter. So on the one hand the sunspots are darker, on the other, these smaller magnetic concentrations are brighter, perversely for the same physical reasons.

These photospheric properties reveal that fluid and magnetism are *strongly interacting*. Other manifestations of these interactions, collectively called 'magnetic activity,' are seen more dramatically above the photosphere, where the plasma is more tenuous, and the magnetic fields have a much stronger influence.

Magnetic forces

As gravity has a field surrounding heavy masses like the Earth, which exerts a familiar force on an object with mass, bar magnets exert forces on objects carrying electric currents. Bar magnets contain many atomic-sized rings of electric current, aligned, an effect called ferro-magnetism. Figure 7 shows patterns of magnetic fields surrounding bar magnets, and illustrates the two kinds of forces between them. When similar poles are near one another, each feels a repulsive force, as if there is a *pressure* in the field lying between. When different poles are adjacent, they experience a force of attraction, as if a strong elastic band was stretched between them. Mathematics confirms that magnetic stresses are a sum of tension and pressure. Tension acts only along magnetic lines of force, pressure in all directions equally.

The poles of the Sun seen during eclipse, shown in Figure 10, trace out similar lines of force to those at the poles at the left and right edges of the figure.

The region immediately above the photosphere is the solar *chromosphere*. The name, 'colour-sphere', comes from a characteristic red hue seen for a few seconds at the beginning and end of totality during a solar eclipse. A British naval sea captain, Stannyan, holidaying in Switzerland, reported in a letter to Astronomer Royal John Flamsteed (1646–1719) a reddish streak around the solar limb for a few seconds just as the Moon uncovered the Sun during the eclipse of 12 May 1706. More than a century later, George Airy (1801–92), then Astronomer Royal, described seeing a 'red sierra' of mountain-like structures above the Moon's dark disc during the eclipse of 1851. But it was Edward

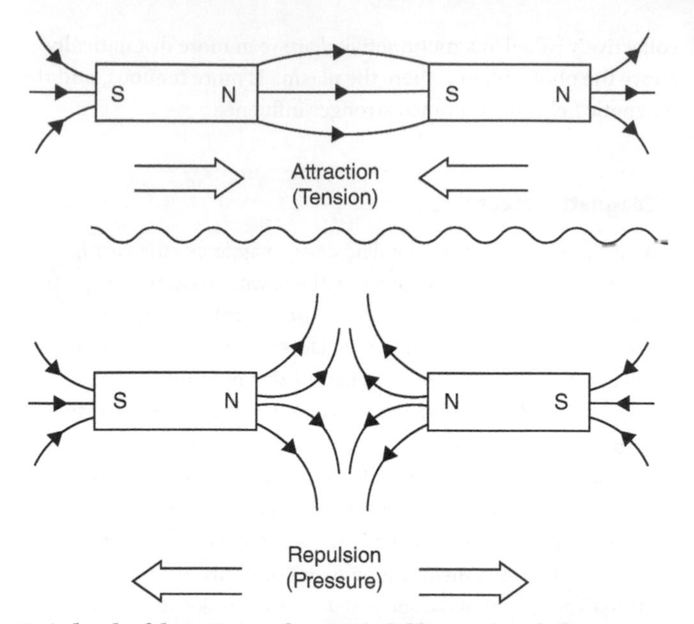

7. A sketch of the patterns of magnetic fields traced out by bar magnets. Two interacting bar magnets show how the magnets contain energy in the fields around them. When similar poles are aligned, there is a repelling force similar to a high pressure between the magnets. When oppositely-aligned, the force is attractive as if there were tensed-up elastic bands pulling the magnets together.

Frankland (1825–99) and Lockyer who, in 1869, first used the term 'chromosphere' for this region. The next year, astronomer Charles Young showed that the chromosphere corresponds to the layer of the Sun's atmosphere responsible for the darkest lines in Fraunhofer's solar spectra. The red colour originates from the brilliance of the Hα line of hydrogen at a wavelength of 656 nano-metres (nm; 10^{-9} metres). Modern instruments can isolate light in wavelength bands narrower than the solar spectral line widths, permitting us to separate chromospheric from photospheric features while observing the bright solar disc. The first instruments to achieve this, spectroheliographs, were developed in the 1890s independently, first by Hale and shortly afterwards by Henri-Alexandre Deslandres (1853–1948).

Eclipse observations revealed that the chromosphere is sandwiched between the dense photosphere and tenuous corona, because chromospheric emission is seen after the photosphere has been covered by the Moon, and before and during the appearance of the corona. Therefore Young's darkest lines must originate above the photosphere and below the corona. Just as the Earth's atmosphere is layered into the troposphere, stratosphere, mesosphere, etc., the horizontally-averaged structure of the Sun's atmosphere is layered into photosphere, chromosphere, and corona. Between the latter is the (unimaginatively-named) transition region. In the Sun's chromosphere and above, the average layering is strongly perturbed by the magnetic fields driven from the dense layers beneath, leading to dynamic phenomena such as spicules, flares, and other curiosities.

Origin of magnetic force and stress

How do forces between bar magnets arise? Experimenting with cathode rays, Joseph John Thomson (1856–1940) discovered the electron. In 1881 he showed that the rays (electrons) were deflected perpendicular both to the direction of the rays and of the magnetic lines of force. As a moon orbits in circles around a planet through the inward pull of a gravity, the perpendicular force leads to circular motion. As the planet moves in space, the moon will move in a helix. Likewise electrons moving along and across lines of force exhibit helical motion. By 1895, Dutchman Hendrik Lorentz (1853–1928) arrived at the formula:

$$\textbf{Lorentz force} = q[\mathbf{vB}].$$

(continued)

Continued

Here, charge q moves at velocity **v** in magnetic field **B**. Bold lettering indicates vectors. The notation **[vB]** means a force proportional to speed v, magnetic field strength B and $\sin \vartheta$, where ϑ is the angle between **v** and **B**. The force acts in just one of two possible directions perpendicular to **v** and **B**, for a given charge. Readers might recall using the 'right-hand rule' at school, to choose the direction that corresponds to nature. It is an important example of the breaking of symmetry. When included in Newton's laws, the equation leads to helical motion.

Inside a bar magnet, atoms form sub-microscopic electrical circuits as electrons orbit their nuclei. The nuclei are essentially fixed in space, but the electrons of a ferromagnet, such as an iron bar, all orbit around the north–south line of the magnet, reinforcing the magnetic fields of one another. Summing all the currents leads to the generation of large-scale magnetic fields around the bar, and the macroscopic forces illustrated in Figure 7. In solar plasmas too, we can frequently ignore ion motion. Averaging electron motions in each cubic metre, the electric current per square metre is $\mathbf{j} = -en\mathbf{v}$, where n is the number of electrons with charge $-e$ per cubic metre, and their average velocity is **v**. The Lorentz force per cubic metre of fluid or plasma is simply

$$\textbf{Lorentz force} = [\mathbf{jB}].$$

The first evidence for this structure in the chromosphere is reproduced in Figure 8, a sketch from visual observations by Jesuit priest Angelo Secchi (1818–78). When observing above the solar disc, there is almost no light in the background. Everything along each line-of-sight that skims above the solar surface becomes visible. Secchi noted the forest of features sticking out above the photosphere. The forest, seen from Earth across the solar limb, appears quite thick. But seen from above, it is clearly quite sparse. These long, thin structures that extend above the stratified

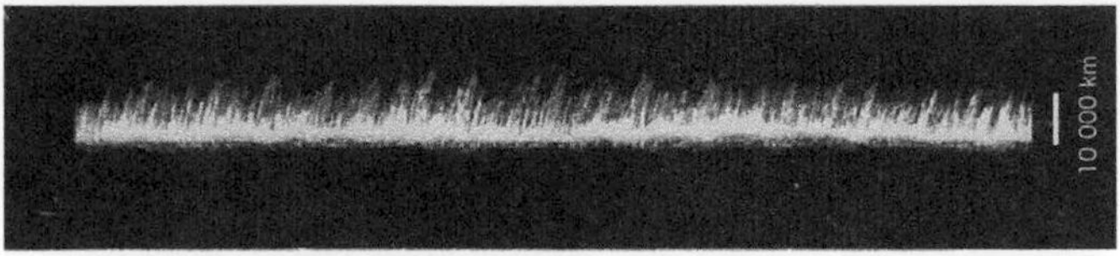

8. **A sketch of the chromosphere seen at the limb of the Sun, made by Father Angelo Secchi in 1869. An approximate scale is annotated at the right. The solar radius is 700,000 km.**

(layered) part of the chromosphere are called 'spicules'. These features originate within the chromosphere. They extend into the hotter corona. Secchi's sketches show incredibly small details, down to 500 km or smaller. It took a century or so to improve on the details seen in these kinds of images, owing to the difficulties in observing the Sun through the Earth's atmosphere. Secchi's success is probably due to our ability to pick out sharp features during occasional moments of extreme clarity that do occur when observing from the ground. It takes elaborate adaptive optical systems and computer algorithms to obtain results of consistently higher quality.

One of the highest quality observations of the photospheric—chromosphere—corona interface is shown in Figure 9. Long dark streaks are visible, some of which are related to spicules seen above the solar limb. In stark contrast to the granules, these streaks and spicules exhibit a great deal of order on scales larger than granules. Thin structures are almost uniformly assumed to trace out magnetic lines of force originating from concentrations of photospheric magnetic fields underneath. These structures seen on the solar disc tend to live considerably longer than spicules seen at the limb. The connection between these features seen at the limb and on the disc is still being actively discussed today.

The chromosphere spans pressures from 100 Newtons per metre $(\mathrm{N/m})^{-2}$ at the top of the photosphere, to 0.02 $\mathrm{N/m}^{-2}$ at the base of the corona (see Figure 4). This drop, a factor of 5,000, occurs in

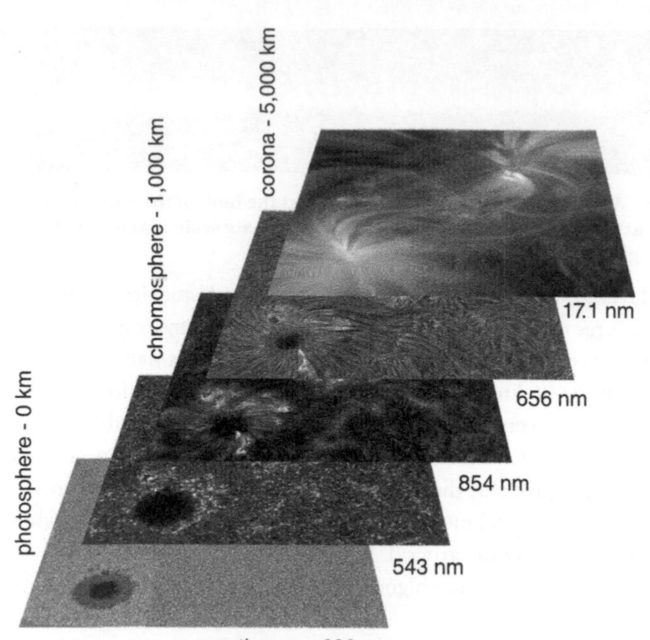

9. Images of the solar atmosphere from photosphere to corona. The labels indicate the wavelength in which each stacked image was obtained, in nano-metres (nm). The topmost image, showing the corona, samples extreme UV wavelengths in a broad band near 17.1 nm. The lowest image is in a broad band near 600 nm, the rest are narrow band images, modern versions of Hale's spectroheliograph image of Figure 22. The 543 and 854 nm images show details of plages, the bright patterns away from the sunspot.

a vertical distance of just 1,500 km, about the diameter of a single granule. The radical drop in pressure is caused by the low plasma temperatures of order 6,000 K or less, in this, the coldest region of the Sun. The spicules extend far higher, typically about 7,000 km. Spicule plasma appears to move at supersonic speeds. If they are highly supersonic jets, drawing energy for acceleration from magnetic fields, their plasma can reach well above the bulk of the chromosphere. But we do not yet have a fully consistent picture of the nature or origin of such long, thin structures that extend into the corona. Spicules may be important in supplying the corona with mass.

Stratification

The different appearance of photosphere and chromosphere reflects differences in the nature of stresses (e.g. pressures) in the two regions. Plasma pressures drop exponentially with increasing height through the photosphere and chromosphere. The same behaviour occurs in our atmosphere. But the magnetic stresses (magnetic pressure and tension) drop far more slowly with height, obeying laws of Maxwell, not those of hydrostatics.

An exponential drop in pressure is familiar to those ascending the highest mountains. At sea level, the air pressure is 100,000 N/m^{-2}, or 14 pounds per square inch. But at 8.8 km (the height of Everest), the air pressure experienced by Edmund Hillary (1919–2008) and Tensing Norgay (1914–86) is three times lower; 8.8 km higher, the pressure drops by another factor of three, and so on. Submariners experience a less rapid pressure change with depth. For every 10 metre increase in water depth, the pressure increases by one additional atmosphere of pressure (i.e. 100,000 N/m^{-2}). Air is compressible, but seawater is (almost) incompressible. Both are examples of fluids in hydrostatic equilibrium, where pressure supports the weight of overlying material. The left-hand panel of Figure 11 illustrates exponential (atmosphere) versus linear (seawater) behaviour. Low in the solar atmosphere, the plasma contains more energy than the magnetic field, and *vice-versa*. As a result, the nature of solar magnetic fields changes from disorder to order, while traversing the the solar chromosphere. The chromosphere is usually the region where energies switch from thermal below to magnetic dominance above.

If doubt were expressed as to the magnetic structuring of the chromosphere, surely all is removed when examining the solar *corona*. Frank H. Bigelow (1851–1924) remarked on the similarity of structures observed in the corona during eclipses to patterns associated with terrestrial magnets (Figure 7). Figure 10 shows a highly processed image of the corona during eclipse. It reveals patterns familiar from iron filings around bar magnets (shown in Figure 7), seen, for example, near 11 and 5 o'clock. The brighter helmet- and arch-like structures contain most of the coronal plasma. Unlike the light from the Sun's disc, we see all the way through the corona. Therefore the figure shows overlapping structures along each line-of-sight, as in Figure 8.

Between the World Wars, enormous progress was made in both theoretical and experimental physics, alongside more modest

10. Image of the solar corona as photographed from Mitchell, Oregon on 21st August 2017 by Miloslav Druckmüller. The image has been processed to highlight structure, the corona is much brighter near the Sun than the figure suggests. Notice the curved 'rays' pointing outwards at 11 and 5 o'clock, the semi-circular features around 4 o'clock, and the three long flame-like structures pointing far from the Sun at 1, 3, and 8 o'clock. The 'rays' are analogous to patterns of iron filings emerging out of the poles of a bar magnet (Figure 7).

advances in solar work. Using data from real and artificial eclipses, made inside 'coronagraphs' developed by Bernard Lyot (1897–1952), in 1939 Walter Grotrian (1890–1954) intimated that the corona may be very hot. During the 19th century, a prominent spectral line in the corona resisted identification in the laboratory. This 'coronal green line' was associated, like the D3 line of helium, with a new element, 'coronium'. By the middle of World War Two, Edlén had demonstrated that the coronal spectral lines belonged to elements such as iron, silicon, and calcium, but *with multiple electrons removed*. The shocking conclusion was unavoidable. Coronal plasma was ionized far more than plasma at the solar surface! Coronal electrons were soon inferred to have temperatures in excess of one million Kelvin. (The Kelvin scale starts at absolute zero, where all thermal motions cease. Simply add 273 to the Celsius scale to obtain temperature in Kelvin.) 'Coronium' was gone. The green line was a transition of an iron ion, with thirteen electrons removed!

The reason for scientists' reticence and initial disbelief of a hot corona lies in the vaunted Second Law of Thermodynamics, which states that heat does not flow from cold places to hot. Yet the corona is much hotter than the underlying atmosphere, so heat must flow from the corona downwards, by the Second Law. Therefore, a more ordered form of energy, not heat, seemed to be heating the corona. Heat is, after all, the lowest form of energy, consisting of the kinetic energy ('KE': energy associated with motion) of random motions of particles or molecules. In an era that witnessed the rapid development of quantum theory of particles and fields, the mundane Sun was delivering a genuine scientific mystery.

In further dramatic contrast with the disorder of photospheric magnetic fields shown in Figure 6, the bar magnet and other fields exhibited by the corona (Figure 10) are orderly. Yet, outside of sunspots which generally lie close to the solar equator, the coronal structures must originate somehow from the disorderly

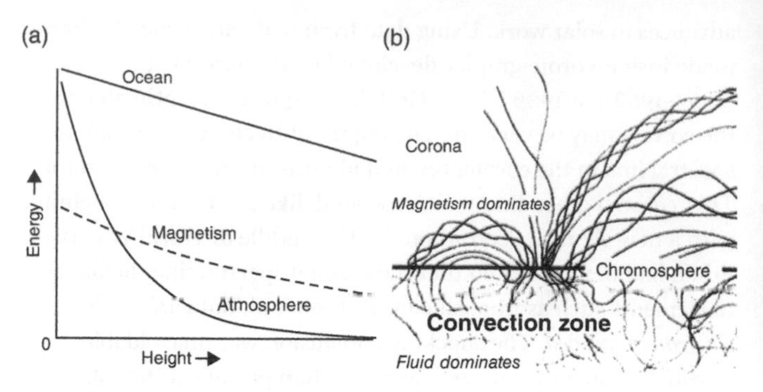

11. Left: A sketch of how energy densities in gas and plasma fall with height (solid lines), and in the magnetic fields (dashed line). Right: A vertical slice through a conceptual sketch of the regions below and above the solar photosphere (immediately below the horizontal line).

photospheric fields. The photospheric fields are truly tiny—organized by convection to scales less than 0.1 per cent of the Sun's radius (Figure 6). Exactly how the fields emerge through the chromosphere, expanding to fill the corona, is a question that remains only partly understood. The chromosphere hosts what we might consider as a *magnetic* transition region, as opposed to the *thermal* transition region separating chromosphere and corona. The magnetic fields below it are forced around by convection, but within and above it the magnetic fields force the plasmas.

Another curiosity was a solar phenomenon that has characteristics of the chromosphere, but existing high within the corona. 'Prominences' were observed using spectroscopes in the 19th century, first during and then outside of eclipses. A famous example is shown in Figure 12, an image obtained at the High Altitude Observatory (HAO) in Climax, Colorado during the first operations of a new coronagraph. The light from these structures is characteristic of the chromosphere. They can extend across a solar radius, and consist of cool material suspended by magnetic forces within the corona. This enormous prominence exhibits

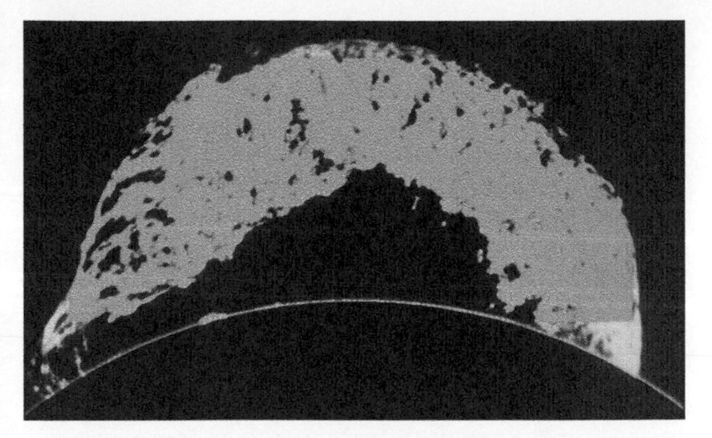

12. One of the first photographs obtained with the new coronagraph of the HAO at Climax, Colorado, on 4 June 1946. The instrument blocks the solar disc (dark region at the bottom) and allows light of hydrogen to enter from the corona. The telescope operator expressed surprise at the large size of the structure seen.

helical structure, an indicator of magnetic forces in action. Given their diversity and obvious complexity, they remain a subject of intense research for their own sake and for understanding physical conditions within the corona.

The orderly patterns of the corona shown in Figure 10, while similar, are not identical to those of iron filings around a bar magnet. In particular, not all features appear to return to the Sun, they appear stretched out into space. Typical examples are the three bright 'flames' at 1, 3, and 8 o'clock in Figure 10. This is because the solar wind stretches the magnetic fields away from the surface. The wind is in part a by-product of mechanisms that heat the corona. Once the corona is hot, the pressure drives plasma outwards unless constrained by magnetic tension forces. Before the 1950s, the observation of comet tails drawn out in directions away from the Sun suggested to some that the Sun exerted two or three kinds of force in interplanetary space. One seemed to affect the neutral gas, another ionized plasma, and a third the comet's 'dust', solid particles released as the comet is heated by solar radiation. But it took two theoretical advances in the 1950s to establish a theoretical explanation for two of the forces. The breakthrough was made by Parker in 1957. He examined the force and thermal balance of the corona. A year earlier, Sydney Chapman (1888–1970) showed that coronal plasma at 1 million K is a very efficient conductor of heat, so that the corona must be hot over a considerable fraction of the inner solar system. Using this result, Parker showed that even a static corona exerts a finite pressure far from the Sun. This pressure could not be balanced by the known pressure of the interstellar medium. Parker therefore concluded that the corona must inevitably *expand*. Parker's paper was met with great scepticism. Then, in January of 1959, direct measurements of particles outside of Earth's magnetic shield (the 'magnetosphere') were made with the Soviet spacecraft Luna 1, results which were soon reproduced by Luna 2. Mariner 2, on its way to Venus in 1962, measured the direction from which these interplanetary particles came, confirming their solar origin and discovering also that this solar wind flowed continuously, but with significant fluctuations. Later we will see that the Sun's wind has a fast and slow component.

Beginning in the 1950s, radio astronomy offered insight into the solar wind by observing the effects of the coronal plasma on radio signals from background sources. Nobel Laureate Anthony Hewish (1924–) pioneered this work, demonstrating how variations in the density of the solar corona translate into scintillations (random rapid fluctuations) of radio sources. The measurements enable one to derive various plasma parameters of the solar wind and corona, a few solar radii above the Sun's surface. In recent decades this has included signals from spacecraft on the other side of the Sun, such as the Ulysses spacecraft. This work will receive more attention as the Parker Solar Probe mission (Figure 40), launched 12 August 2018, dives deep into the solar wind over the next decade.

In the era of space exploration our understanding of the corona and wind has steadily increased. Space allows us to 'see' wavelengths below the atmospheric cut-off wavelength of 310 nm that are blocked by our atmosphere from reaching the ground. In X-rays (wavelengths below about 10 nm), the solar photosphere and chromosphere appear dark because the temperatures there are just too low. But the hot corona radiates UV- and X-rays, and it is readily seen against the dark disc of the Sun. The advent of the first long-lived X-ray solar observatory on board SKYLAB led to

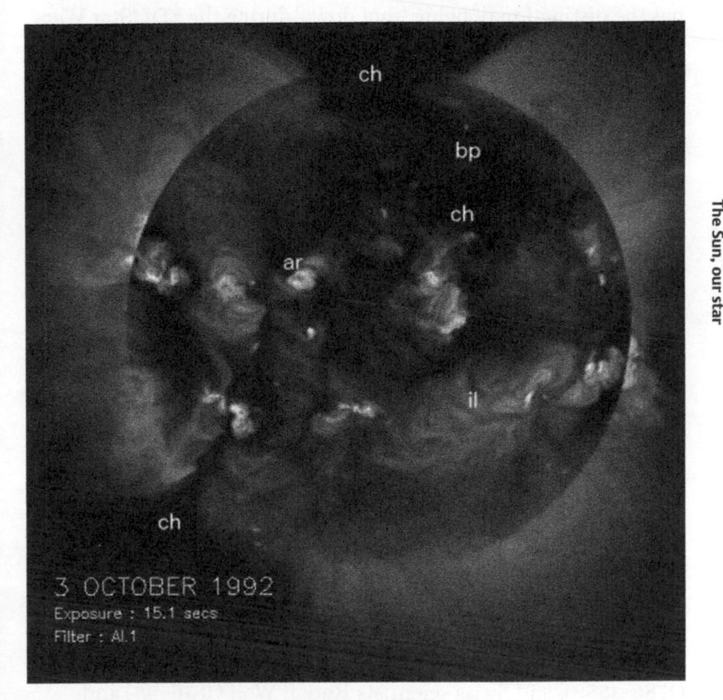

13. The whole corona seen in X-rays with the Yohkoh spacecraft, showing plasma near two million Kelvin. Small 'bright points' ('bp'), active region loops ('ar'), loops connecting different active regions ('il'), polar and low-latitude coronal holes ('ch') are all visible. Magnetic fields organize all the structure visible in this picture.

many discoveries. Among the new phenomena, observers identified striking and persistent dark regions in the corona, large regions now called 'coronal holes'. An example is shown in Figure 13. Such holes are the source of the fast component of the solar wind.

A third type of dynamical structure sometimes makes abrupt appearances in interplanetary and Earth space. They were originally identified as a separate phenomenon in 1979, using the NRL SOLWIND coronagraph on board the P78-1 satellite. (P78-1 has the distinction of being 'shot down' during the SDI 'Star Wars' initiative.) These so-called CMEs not only eject mass in the form of plasma, but also they drag magnetic fields into interplanetary space. CMEs can seriously disrupt our telecommunications systems, and even our electrical power supplies, as well as generate spectacular auroral displays. The CMEs are most readily observed from space because of the brightness of the daytime sky and relative dimness of the CMEs. An example is shown in Figure 29 in Chapter 4.

Sun–Earth connections

The most obvious link between Sun and Earth is the near-constancy of the Sun's radiant energy, enabling the processes of chemistry and then biology to produce complex organisms over several billion years. But, as we have seen, the Sun also affects the Earth in ways unanticipated from first principles, through the emission of high energy radiation, particles, and magnetic fields, and abruptly through CMEs. A connection between the Sun and well-known perturbations to the Earth's magnetic fields was suggested at least as far back as 1852, when Edward Sabine (1788–1883), and Swiss scientists Rudolf Wolf (1816–93) and Alfred Gautier (1793–1881) independently discovered a correlation between spots on the Sun, geomagnetic fluctuations, and occurrences of the Aurora Borealis. In 1872, Young noted that some perturbations were preceded by flaring on the Sun.

He highlighted what is now called 'The Carrington Event', the first report of a flare seen on the Sun by observers Richard C. Carrington (1826–75) and, independently, Richard Hodgson (1804–72). Young and others pointed out connections between this flare and subsequent disturbances to the Earth's magnetic field shortly afterwards, including spectacular aurorae. Edward Walter Maunder (1851–1928) quantified relationships known for some fifty years between the waxing and waning sunspots, and geomagnetic activity, in 1904.

Kristian Birkeland (1867–1917) led Arctic campaigns to observe geomagnetic fields and the Aurora Borealis, often under brutal weather conditions. Together with laboratory experiments, in his 1908 book he offered a physical picture, essentially valid today, of geomagnetic variations. Based upon work of Biot and Savart, he postulated that variable electrical currents above Earth's surface cause magnetic variations. His work is celebrated on the Norwegian 200 Kroner bank note (Figure 14).

Maunder had reported a correlation between geomagnetic disturbances and the apparent solar rotation rate of twenty-seven days. He concluded that

14. Kristian Birkeland (1867–1917) is commemorated on the Norwegian 200 Kroner bank note.

> magnetic disturbances have their origin in the Sun. The solar action which gives rise to them does not act equally in all directions, but along narrow, well defined streams, not necessarily radial.

Julius Bartels (1899–1964) related these 'actions' to streams of particles from mysterious 'M-regions' on the Sun in 1932. But it was not until 1974 that the M-regions were identified, using SKYLAB observations, as coronal holes.

In parallel to studies of solar particulate and magnetic influences, the effects of solar ionizing radiation were inferred from the discovery of the Earth's ionosphere. Gugliemo Marconi (1874–1937) famously demonstrated the propagation of radio waves across the Atlantic Ocean in 1901. To explain Marconi's and other observations, in 1902 Oliver Heaviside (1850–1925) proposed a layer in the Earth's atmosphere that can reflect radio waves, named the 'Heaviside-Kennelly' layer, after additional properties of this layer were discovered by Arthur Kennelly (1861–1907) and reported the same year. Edward Appleton (1892–1965) proved the existence of such a layer in 1927, for which he received the Nobel Prize in physics in 1947.

The name 'ionosphere' was first coined by Robert Watson-Watt (1892–1973) in 1926, later known for his contributions to the development of radar prior to the Battle of Britain. The solar origin of the ionosphere became clear from the known night–day variations in radio communications. Quantum mechanics (the photo-electric effect) required that UV and higher energy radiation was needed to ionize the upper atmosphere. Only the Sun could be the UV source, albeit at levels much weaker than the radiation at visible wavelengths (Figure 3). Many researchers have contributed to our understanding of the ionosphere. But special contributions to the basic physics underlying Marconi's original reports were made by Vitaly Ginzburg (1916–2009) during World War Two. Ginzburg was one of the fathers of the nuclear programme of the USSR as well as 2003 Nobel Laureate in physics, for work in superconductivity.

Our understanding of the influence of the Sun on interplanetary space has been greatly increased by several spacecraft: the Solar Maximum Mission (SMM), Solar and Heliospheric Observatory (SoHO), Transition Region and Coronal Explorer (TRACE), and the Solar Dynamics Observatory (SDO) missions. Modern society, dependent on power grids, radio communications, and fleets of satellites, is vulnerable to all these variations in the solar plasmas as they impact the Earth. Unlike earlier generations who had only to deal with weather and adapt to changing climate, our modern society must now also deal with 'space weather', and 'space climate'.

Chapter 2
The Sun's life-cycle

Rising from the ashes

The Sun was born out of the debris of stars which formed and exploded early in the life of our own Galaxy, the Milky Way. Once matter condensed out of the raw energy of the early Universe, gravity determined the fate of that matter. This wholly attractive force dominates across astronomical distances, even though it is the weakest of the four forces of nature: gravity, the weak nuclear force, electromagnetism, and the strong nuclear force.

A major clue to the origin of the Sun and the solar system is encoded in the distribution of the abundances of elements. Through long-established analytical techniques in physics and chemistry, we know the distribution of elements in terrestrial and, from meteorites, some of the interplanetary matter. Before the 1920s it was generally believed that the Sun would have essentially the same composition as the Earth. In the Earth and other rocky solar system bodies, we find little hydrogen and helium. Table 2 compiles abundances of the Sun and in Earth's mantle, highlighting the basic difference.

However, Payne-Gaposhkin's 1925 thesis work implied that hydrogen (H) and helium (He) dominated the composition of the Sun and the majority of stars. The very different terrestrial and solar abundances were understood through the work of James

Table 2. Relative solar and terrestrial element abundances

Element	Solar abundance	Crustal abundance
	(by number %)	
H	93	3
He	7	3×10^{-6}
C	0.03	0.3
N	0.009	0.14
O	0.07	60
Ne	0.007	3×10^{-7}
Si	0.004	20
Fe	0.003	2.3

Notes: H=hydrogen, He=helium, C=carbon, N=nitrogen, O=oxygen, Ne=neon, Si=silicon, Fe=iron. Data are taken from webelements.com

Jeans (1877–1946) and others. The formation of stars and planets is a messy business. The different parts of a condensing cloud of stellar debris gain speed under the mutual force of gravity, but most of the KE (energy of motion) gained is released as radiant heat, as the bits and pieces collide and grind against each other. Hydrogen and helium are the lightest known atoms. When stars form they attract all the mass around them, forming a deep well of gravity out of which most of the stars' atoms cannot readily escape. During the planetary formation process, the gravity is weaker because the condensed mass is smaller. In this environment, some of the lightest atoms have enough energy to escape, their thermal speed exceeding the speed needed to escape from the proto-planet's gravity. In small proto-planets warmed by solar radiation in the inner solar system, most of the hydrogen and helium could escape, leaving only heavier elements behind. Over billions of years, the planetary, cometary and meteoric abundances of light and chemically inert elements became small. In our region of the Galaxy, elemental abundances of stars and interstellar material are much closer to those of Payne-Gaposhkin than those measured on the Earth.

The mere existence of elements heavier than helium is a fingerprint of nuclear processing *after* the Big Bang. In the 'standard model' of cosmology, the Big Bang created only hydrogen, helium, and just one lithium atom in 10^{10} hydrogen atoms. By the 1950s, nuclear physics had advanced enough to identify conditions needed to synthesize elements such as carbon (C), nitrogen (N), oxygen (O), phosphorus (P), and sulphur (S), that with hydrogen form the building blocks for life. It was evident that the interiors of stars, originally formed by gravity of just hydrogen and helium, were a natural environment to synthesize these heavier elements on the main sequence.

Stars can form in a wide range of masses. More massive stars process their hydrogen and helium fuel faster. The earliest of them exploded violently, as supernovae, spilling their processed material, replete with heavier elements, into space. We, along with the Sun, are made of the debris of the Big Bang and such stars, in an ordinary spiral arm of an ordinary galaxy.

Nuclear fusion in nature occurs only in elements with atomic numbers (number of protons in the nucleus) up to 26, which is the element iron (Fe). In laboratories, one can fuse heavier elements but such processes require energy beyond what is available under natural conditions. So where did these still heavier, and rarer, elements with atomic numbers from 27 to 92 (uranium) originate? Laboratory nuclear physics tells us that elements such as platinum, with atomic number 78, are formed by the exposure of iron nuclei to a large flux of neutrons. These neutrons are released in cosmic explosions such as supernovae, and in the interiors of giant stars. Such elements are therefore rarer than most lighter elements. Their paucity is reflected in marketplaces throughout history.

The spiral arms of galaxies are full of this enriched gas, plasma, and dust, most obviously seen in 'interstellar clouds'. A well-known example is the Orion Nebula visible to the naked eye, just below Orion's belt. The theory of stellar evolution and element nucleosynthesis is one of the most complete and successful intellectual achievements of 20th-century science. We can be confident in our understanding of the environment in which the solar system and the Sun were born.

Star formation

But how exactly did the Sun coagulate out of the chaotic ashes of supernovae? In short, because of gravity, and the natural variations in density, the supernova debris that comprises an interstellar cloud will tend to clump. The clumps grow and form stars, planets, and other objects. The manner in which clumping occurs is interesting. The process involves the formation of a disc structure. Not only does a disc permit the star to form, but it naturally forms the system of planets, comets, and related bodies. The Sun rotates on an axis that is inclined at just 7 degrees of the axis of the orbit of the Earth around the Sun, and the Earth's orbit lies in almost the same plane as all the other planets. These orbits all occur within a fairly narrow disc.

Figure 15 shows the environment of a forming star and a close-up of the 'solar system' associated with it. While the Sun is about 4.5 billion years old, this star, HL Tauri, is aged between 0.1 and one million years. The discs shown in the image are up to 4,000 times 1 astronomical unit (AU) in size (1 AU is the distance of 150 million km, between the Earth and the Sun). In our system, these distances happen to lie between the orbits of two sets of rocky objects in our solar system: those in the Kuiper Belt (between 20 and 50 AU, also containing Neptune and Pluto), and others in the Oort cloud (50,000–200,000 AU). The core star of HL Tauri is obscured by dust at visible wavelengths. The Atacama Large Millimetre Array (ALMA) data show 'heat' by observing at wavelengths of a fraction of a millimetre (also called far-IR).

HL Tauri is an object that radiates heat and light through the release of gravitational energy. The mechanism is straightforward. Imagine trying to disassemble the star—run the clock back in

Stellar inflows and outflows

Why must a disc structure form as matter is attracted to form the proto-star? The answer lies in Newton's laws of motion. Just as we define momentum as a product of velocity and mass for a moving body, we define angular momentum as the product of mass times velocity around a given axis, times distance from the axis. A spinning ice skater has a lot of angular momentum as parts of her body speed around her axis of rotation. But she does not move across the ice. Newton's laws imply the conservation of angular momentum. In the absence of friction the skater would spin forever.

A proto-star pulls all matter in a cloud under its gravitational influence towards the centre-of-mass (CM). The cloud must conserve its total angular momentum. Only a tiny fraction of the cloud will have no angular momentum relative to the CM, this fraction being the matter that happens to move *through* the CM. Instead, most mass orbits *around* this centre, thereby carrying angular momentum. In time, the orbiting matter collides with other matter in different orbits, a 'friction' converting KE to heat, and changing the orbits of the colliding matter. But this process *preserves the cloud's total angular momentum*. Those orbits which cause matter to collide frequently do not survive, they are 'filtered out' by collisions. Eventually, the matter spins in an orderly fashion around the CM in a plane, like the horses on a carousel. The end result is a star and a disc spinning with the same total angular momentum that it had before the collapse.

Stellar *outflows* are more spherical because the outflow originates from the star's surface, and is directed largely in the radial direction. They leave the star quickly, minimizing the time for friction to act.

time. You would have to lift the parts of the star away from the accumulated mass, much like lifting a bucket of water. You have to work hard, that is, expend your own energy. In dropping the bucket, equivalent to re-assembling the star, energy is converted to KE as gravity does its work. The speed of the stellar matter increases as it falls towards the proto-star. But the environment does not permit the matter to fall without colliding and mixing with all the other matter. Random collisions convert KE into heat. This is none other than friction which we experience in daily life, for example when we rub sandpaper against wood. The heated matter then radiates, largely at IR wavelengths.

Our detailed understanding of star formation remains incomplete, and this remains an area for active research. Collisions are not sufficient to move all the angular momentum around in the system fast enough to permit the matter to fall efficiently on to the star. Given that the leftover disc material is the stuff of planet-building, this is an important problem, particularly today with the ongoing discoveries of thousands of planets outside the solar system. Here we encounter a potentially critical role for magnetic fields in stars, tied up in the proto-stellar and proto-planetary matter. Magnetic fields can rapidly transport angular momentum, by applying torque from one part of a disc to another through something called the 'magneto-rotational instability'. (Just as forces applied to a particle cause changes in the linear momentum of the particle, *torque*—force times the nearest distance of the force vector from the CM—changes an orbiting particle's *angular* momentum).

Magnetism can play another important role. As gravity pulls the matter together, even a modest fraction of particles that are charged (electrons and protons) experience forces from the ambient magnetic fields. The fields themselves originated via a dynamo process operating within the Galaxy long before. These magnetic fields exert a weak but systematic force on the charges. The charges occasionally impact electrically neutral particles

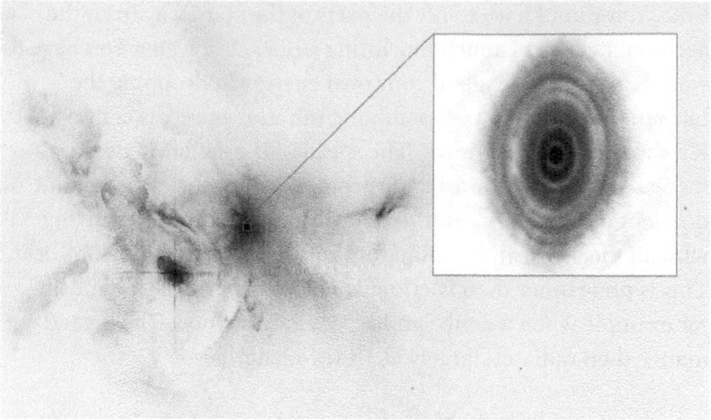

15. A star-forming region in the constellation Taurus is shown as a negative (dark in the image means bright on the sky). The glowing gas in the main image, composed of the ejecta of long-dead stars, is luminous because of the light of several new-born stars. The inset shows an image from the ALMA observatory, showing radiant heat generated around the site of the newly formed star, HL Tauri. This 'heat map' reveals the structure of the matter accumulating around the star. Similar to the rings of Saturn, the structure is a series of discs, some of which may coagulate into planets, in time.

comprising the bulk of the matter, exchanging momentum. The magnetic fields therefore exert a net force on *all* the collapsing matter. Eventually, the magnetic fields are stretched and twisted by motions induced by gravity, often opposing the gravitational forces and so converting gravitational into magnetic energy. In the 1950s, this was a serious theoretical hurdle to overcome. It was as if magnetism were a dark energy pushing against the collapse, a little similar to proposals in cosmological models today, but rooted in known physics. Thomas Cowling (1906–90), in the mid-1950s, showed how collisions between charged and neutral particles can allow the matter to slip across the magnetic fields, permitting the earliest stages of collapse to continue. Modern computer models have revealed that the combination of gravity, magnetic fields, turbulent motion, and the generation of 'jets' of material aligned along the axis of rotation of a proto-star all play a part in star formation. In this sense, our entire existence seems to rely not only on the early history of the Universe, but also on the interaction of (proto-)stellar material and magnetic fields.

The Sun, along with all stars, has gone through a phase like HL Tauri. There was a time when astronomers believed that the Sun radiated precisely due to the release of gravitational energy and friction. But by the 1920s, it was known that gravity could sustain the Sun's luminosity only for about thirty million years. The Earth was then estimated to be ten to one hundred times older. Sir Arthur Eddington (1882–1944) in 1920 suggested that the stars, through most of their lives, must be fuelled through *nuclear fusion*, building on the results of Einstein and Francis Aston (1877–1945). Einstein had showed the equivalence of mass and energy, $E = mc^2$ (energy equals the mass of an object multiplied by the speed of light squared). Aston determined that the mass of a helium atom was 0.8 per cent less than four individual hydrogen atoms. Thus, if a process could convert hydrogen to helium, argued Eddington, it would release the energy associated with the missing mass needed to account for the total energy released by a star over its lifetime.

Once the newly-formed Sun has accumulated most of the mass inside the volume where its own gravity pulls material inwards, it begins a stable period of radiating the gravitational energy released as the star contracts further. This early Sun emits more radiation at a lower temperature than today, as a 'pre-main-sequence star'. This period lasts up to thirty million years, a crude upper limit obtained by dividing the stored gravitational energy of today's Sun by its luminous power. As the star contracts further, the central pressures rise to support the weight of the star. The temperatures increase as the internal stellar material is compressed, just as a bicycle pump increases the temperature of compressed air, even though the star continues to radiate energy to space. Once the core temperature reaches a certain level, the KE of the protons becomes sufficient to overcome the energy of repulsion between protons due to their electrical charge. At that point the strong nuclear force attracts protons together to form helium atoms, most with two protons and two neutrons, though not quite in the manner originally suggested by Eddington.

The main sequence

Today we identify the 'main sequence' physically as the manifestation of the process that fuses light nuclei into heavier nuclei in the cores of stars. This identification is based upon decades of work since the 1920s, in nuclear and atomic physics building upon quantum mechanics, and in advances in astronomy as our ability to measure distances and perform quantitative spectroscopy has improved. But the main sequence was originally identified in observations of clusters of stars—groups of stars roughly at the same distance from Earth—by Ejnar Hertzsprung (1873–1967). In 1907 he plotted the brightness of each cluster star as a function of its colour. Most stars fell roughly along a line: brighter stars were bluer. There was also a clear but less-populated group of red stars that were bright. The latter fall on a different branch, the 'giant branch' in the colour–brightness plots.

Our understanding of the Sun's evolution depends on a link between the physics of an individual star, such as the Sun, and observations of many stars. To see this, consider taking a snapshot of visitors in an art gallery. We would find them mostly in front of the artwork. We would conclude that each individual person spends most time looking at pictures, even though we have just one snapshot of a group. While this seems obvious, the underlying thesis is important. By taking a snapshot of groups of stars, say in a cluster, we find most stars on the main sequence. Therefore the main sequence is the longest phase in the lives of most individual stars. This is an example of the 'ergodic hypothesis' often used in computation, for problems such as climate prediction. It states that the average behaviour of a dynamical system over long times gives the same result as many different realizations of the system seen at one time.

The main sequence is the diagonal strip (top left to bottom right), where most stars are found in the 'Hertzsprung-Russell [HR] diagram' shown in Figure 16. The name acknowledges the seminal contributions of Russell and Hertzsprung. This HR diagram plots physical properties of stars, luminosity (total emitted power) against effective temperature, not the measured properties of relative brightness versus colour, as were first plotted by Hertzsprung. The effective temperature is almost the same as the

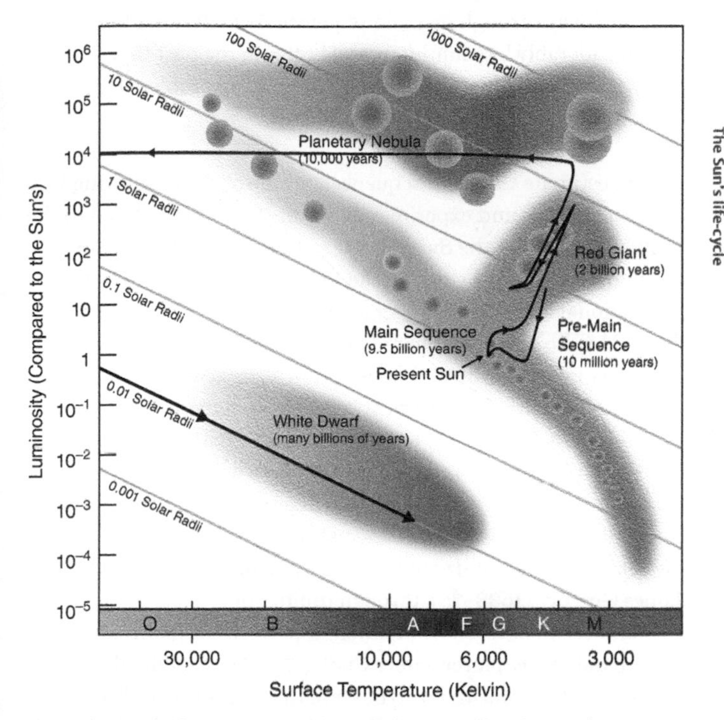

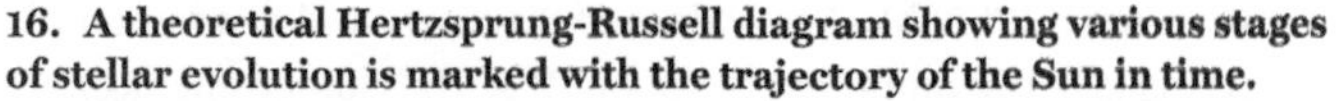
16. A theoretical Hertzsprung-Russell diagram showing various stages of stellar evolution is marked with the trajectory of the Sun in time.

visible surface temperature for most stars. It is defined as the temperature of a black body that emits the same radiative power per unit area as the star itself (see Figure 3). The two kinds of HR diagrams are almost equivalent, intimately related, and form a cornerstone of stellar evolution theory. Since the early work, astronomers have therefore devised ways to compare the two HR diagrams with great precision. Stars arrive on the main sequence complete with a set of element abundances, a certain mass, temperature, and luminosity. While on this sequence, they move slowly to the right and upwards, ending about twice as bright as when they arrived there. Thus the main sequence has a natural width on the HR diagram. At an age of 4.5 billion years, the Sun is about halfway through its main sequence. We infer that the Sun was considerably less luminous and hotter in the past than it is today, which will present us with an interesting problem (the 'Faint Young Sun paradox') discussed later.

We can estimate the main sequence lifetime of a star knowing just a few theoretical and experimental results that were at hand in the 1920s. The mass of the Sun relative to Earth was estimated by Newton as part of his work on the *Principia*, assuming that the gravitational force on an object is the product of the two masses, divided by the distance squared between them, times a fundamental constant. This assumption was necessary for compatibility with the empirical laws of planetary motion established by Kepler (1571–1630). In 1798, Henry Cavendish (1731–1810), using a torsional balance, reported the density and mass of Earth. In so doing he had to determine the gravitational constant along the way. He made use of earlier landmark measurements of the size of the solar system, based on an idea of James Gregory (1638–75), through timings of transits of Venus of 1761 and 1769 observed from different parts of the globe, and using parallax. Explorer James Cook (1727–79) captained a voyage of HMS *Endeavour* to observe the Venus transit from Tahiti in April 1769. Modern values of distances and fundamental constants give the Sun's mass as about 2×10^{30} kilograms (kg).

Armed with Aston's mass differences between four protons and helium, Einstein's $E = mc^2$ relation, Eddington's fusion hypothesis, and the measured rate of energy loss from the Sun today, we find the Sun loses—by radiative energy—about four thousand million kilograms per second. As measured near Earth, the solar wind loses between one and two thousand million kilograms per second. If we allow just 1 per cent of all the hydrogen in the Sun to be converted into helium (in the deep interior where the reactions can take place), we find it can continue to do this for 160×10^9 years, some 5,000 times longer than can be sustained by gravitational collapse. The age of the Universe is currently estimated to be 13.8×10^9 years.

Eddington's initial idea of combining four protons and two (far lighter) electrons to form helium nuclei faced challenges from experimental nuclear physics performed in 1919 at the Cavendish Laboratory. The energies of the core protons, as characterized by Eddington's core temperatures from classical thermal physics, were insufficient to allow the needed reactions to occur against the electrical repulsion between protons. In 1928, during the rapid development of quantum mechanics, George Gamow (1904–68) showed how some particles can penetrate nuclei at lower energies ('quantum tunnelling'). A year later Robert Atkinson (1898–1982) and Fritz Houterman (1903–66) applied these ideas to stars. The problem remained a subject of intense interest when, in 1938, Gamow organized the Fourth Washington Conference on Theoretical Physics. Edward Teller (1908–2003), his student Charles Critchfield (1910–94), Hans Bethe (1906–2005), and Gamow were among those determined to solve the stellar energy problem. Critchfield summarized ideas on *chains* of reactions involving collisions between just two elementary particles and/or radioactive decay. Bethe soon concluded that Critchfield's ideas for helium fusion were sound, and further proposed carbon as a catalyst that might provide an alternate chain of reactions for more massive stars than the Sun.

The two chains are called the 'proton–proton cycle' and the 'CNO cycle'. Today they form the basis of our understanding of how stars evolve.

In fact, in a scientific twist which might have delighted Sir Arthur Conan Doyle (1859–1930), the theory has survived intact to the extent that it has revealed problems in particle physics. The 'solar neutrino problem' was shifted to a problem in elementary physics. The story is reminiscent of work by Fred Hoyle (1915–2001) in the 1940s. Hoyle was known for his remarkable breadth, including a famous 1957 work addressing the origin of elements heavier than iron. He famously predicted specific properties of a nuclear reaction (now called the 'triple-alpha process') based upon the idea that observed stellar properties could be trusted enough to infer properties of the physics of elementary processes between particles.

First predicted by Wolfgang Pauli in 1931 to explain properties of observed products of nuclear β-decay governed by the weak nuclear force, neutrinos were only found experimentally in the 1950s. The 'solar neutrino problem' arose through advances made later: Raymond Davis (1914–2006), along with John Bahcall (1934–2005), had pushed to measure these ghostly, weakly-interacting particles, to test the chain theory. By 1970, Davis and colleagues had accumulated evidence sufficient to reveal a significant problem. Their detector was based upon a large tank of dry-cleaning fluid in the Homestake Mine deep underneath South Dakota. Later work confirmed a dearth of neutrinos generated in the proton–proton chain in the Sun. There were at most half as many neutrinos as predicted by theory. The Homestake results were confirmed in the late 1980s independently using a different detector high in the Japanese Alps, called Kamiokande II. The experiments consistently showed neutrino counts 1/3 of those predicted by the solar (interior) 'standard model' carefully assembled by Bahcall. These results were disturbing, but at the same time new measurements of the solar interior structure were beginning to emerge from an entirely different direction.

Just as seismic waves generated by earthquakes and measured at the surface allow us to understand the internal structure of the Earth, the Sun also supports wave motions in the interior that carry information on conditions there. By 1974, Franz-Ludwig Deubner (1934–) had identified that the surface motions contain information on wave motions, like those of a bell, that propagate through the interior. By 1987, analysis of these oscillations revealed quantitative information on the global interior conditions in the Sun that affect the speed of sound. This speed depends on the internal temperature and is affected by ordered motions of the interior fluid, such as rotation, and by magnetism. These and subsequent measurements confirmed, to the degree of accuracy required, the thermal structure of Bahcall's standard model. The factor of three discrepancy was satisfactorily explained only after it was found, following a suggestion by Bruno Pontecorvo (1913–93) in 1969, that neutrinos in fact possess a small mass. This opened up the possibility for neutrinos to change their type ('flavour'), and hence, alter their detectability. By 1999, the Sudbury Neutrino Observatory produced definitive evidence of neutrinos changing their flavour, and after extensive work it was found that only 34 per cent of electron neutrinos from the Sun would be detected by the Homestake experiment.

In a real sense, the stellar main sequence observed in galaxies across the Universe is the result of nuclear fusion. In this way the large-scale Universe reflects nature at its smallest measurable scales. Without this long-lived, stable phase in stars, complex life could not have evolved on Earth. The fossil records, as well as the imprints left by processes affecting different atomic isotopes, are a testament to the relative stability of the Sun over its 4.6 thousand million year life on the main sequence. We owe our existence to many properties of the Universe, one of which is the simple notion that stars are long-lived on a main sequence defined by nuclear physical processes on sub-microscopic scales. I find the connection between sub-microscopic and cosmic scales to be a pleasing success of modern physics.

Angular momentum and magnetic activity

So far we have discussed the thermal evolution of the Sun. But each star is endowed with a non-zero angular momentum resulting from the natural asymmetry of the star-forming process. Today the Sun spins roughly once every twenty-seven days. The angular momentum of an isolated object is not easy to shed, so this remnant spin should be no surprise. But this rotation is just an average value. Not being a solid, the Sun rotates faster at the equator (once every 24.5 days) than close to the poles (about thirty-seven days). We say that the Sun exhibits *differential rotation* with latitude, at the solar surface. The fluids at neighbouring latitudes move relative to one another, in a 'shearing' motion. It is not the dominant motion, but it is ordered on large scales, unlike the smaller-scale convective motions that are much more energetic and random, as seen in Figure 6. Like the large ocean current systems such as the North Atlantic Drift, a swimmer would not feel this current compared to the local wave motions. But the British Isles certainly are influenced by this sedate, but steady, drift.

The Sun did not always spin so sedately on the main sequence. We know this for two reasons. First, using the ergodic hypothesis, we can observe stars in our Galaxy at all different evolutionary stages. By identifying Sun-like stars at various ages (a challenge requiring some ingenuity), we can place stars on the solar trajectory shown in Figure 16, and imagine the Sun progressing along this track in time. Young stars spin faster than middle-aged stars. Second, since the late 1950s, we have known that stars possess winds—some of the mass is shed back to the interstellar medium during pre-main sequence and main sequence evolution. This form of loss of mass is different from the mass loss due to radiation (energy) losses alone. The ejection of mass from a rotating star is accompanied by angular momentum loss. Newton's third law says that to every action there is an equal and opposite reaction. In effect, the mass in the solar wind is pushed out by some process(es) but because the Sun is rotating it endows the ejected material with angular momentum. The total angular momentum is conserved, so the Sun must lose angular momentum. Both observations and theory imply that the Sun used to spin faster in the past.

The subject of angular momentum evolution began in earnest in 1972 when Andrew Skumanich (1930–) published a two-page paper relating stellar surface rotation rates to stellar ages. Skumanich found that rotation rates scaled as the inverse square root of age for a sample of main sequence stars. The implications were clear, the stars were losing angular momentum in a systematic fashion, via a stellar wind, no other explanations being viable. Skumanich also studied the brightness of a chromospheric line of singly ionized calcium (Ca^+) that was shown, in 1959, by physicist R. B. Leighton (1919–97) to be tightly related to solar surface magnetism. As the Ca^+ data followed the same inverse square root relation, all three quantities appeared to be strongly related: rotation, chromospheric emission, and magnetism.

In recent years, analysis of rotation rates of older Sun-like stars suggests that the Sun may not spin down any more, until it leaves the main sequence. Nevertheless, in one short paper, Skumanich set the observational basis for understanding the evolution of stellar angular momentum and magnetism. We will study further consequences later, when we discuss the young Sun, magnetic effects on the Earth, and large-scale solar dynamos.

After the main sequence

We believe that the evolution of stars is sufficiently well-understood that we can predict the future of the Sun and solar system with some precision. In six thousand million years the

Sun will make a graceful exit from the main sequence. It will move to the right (cool) and upwards (brighten) in the HR diagram, and begin to ascend almost vertically, for about 1 per cent of its main sequence lifetime. The evolving Sun will then have dramatic impacts on the solar system. The manner in which the Sun evolves is remarkable, again combining physics on large and sub-microscopic scales.

Models of future solar evolution differ from those on the main sequence in that the sub-microscopic effects of quantum mechanics become important on global scales. The exit from the main sequence begins when the stellar core consists only of the ashes of fusion of hydrogen, that is, helium. But the core temperatures are insufficient to fuse helium. With no more energy generation in the core, it contracts under gravity, and overlying fresh hydrogen from the radiative zone moves inwards. Fusion of hydrogen to helium then restarts in a shell around the inert core, as temperatures rise because of the heat generated by the continued release of gravitational energy. At the beginning of this 'red giant' phase, shell burning releases several times more power than the core main-sequence phase. In response the outer layers of the Sun expand greatly, the photosphere possibly reaching one hundred times its current radius. The Sun will then radiate at surface temperatures below 4,500 K, dropping to near 3,000 K at the very top of the giant branch, peaking in the red region of the spectrum (see Figure 16).

The benign burning hydrogen shell steadily adds mass to the core, until enough weight builds up that the core can no longer be supported by thermal pressure. There follows a catastrophic core collapse that is only stopped by a stellar-sized manifestation of quantum mechanics.

Unlike the collapse of matter under gravity to form a star, or a galaxy, atoms do not collapse under the force of electrical attraction between protons and electrons because of one of the principal effects of quantum mechanics. The electrons belong to a class of fundamental particles called *Fermions* after Nobel Laureate Enrico Fermi (1901–54), which are characterized by a spin quantum number of 1/2. They obey the exclusion principle, first invoked by Pauli to explain the shell structure behind the periodic table of the elements, among other chemical properties. Pauli's principle, later found to have its root in the fact that elementary particles are not distinguishable and so must obey particular symmetries, requires the electrons surrounding the nuclei of elements to occupy separate states. The six electrons of a carbon atom, for example, do not simply fall into the carbon nucleus, but instead, each electron must occupy its own specific place in three-dimensional space, also with its own definite spin angular momentum. The electrons cannot simply pile into one state, like planets aligned along the same classical orbit. This entirely quantum phenomenon allows electron shells to build up around nuclei, conferring on each atom its unique chemical properties. It also forces atoms to occupy a much larger volume than would otherwise be the case.

In the same way, the collapsing stellar core cannot force all the electrons within it into a space smaller than the quantum physics allows. The core is supported entirely by these quantum conditions that amount, macroscopically, to an enormous force between the electrons. The ions are stuck with these 'degenerate' electrons because of the large electrostatic forces. The collapse is accompanied by increasing temperatures, but with no significant expansion. This is because ordinary pressure (due to heat) is small compared to the 'degeneracy pressure' (force per unit area) of the tightly-packed electrons.

When the core temperatures approach a hundred million Kelvin (six times higher than on the main sequence), fusion of helium nuclei occurs with the release of enormous amounts of energy. This leads to a 'helium flash', an extremely fast runaway explosion, as helium suddenly fuses into carbon, with a small amount of oxygen, throughout the core. The energy release reconfigures the entire star, the extra energy per particle creating an enormous thermal pressure, lifting the core electron degeneracy. The star readjusts to make an unobservably rapid return near to the bottom of the original red giant phase (again see the track in Figure 16). The Sun will live for some time in this 'clump', a phase of non-degenerate core helium burning. This might be considered as a kind of helium-burning main sequence. The equivalent of the proton–proton chain is the triple-alpha process, in which collisions between helium nuclei form a nucleus of beryllium, and a helium–beryllium collision forms carbon, accompanied by a relatively large release of energy again from the higher combined masses of the fusion ingredients and products. As a by-product of the triple-alpha process, some oxygen is also created during the clump phase.

After all the core helium fuel becomes exhausted, another, second ascent of the giant branch in the HR diagram begins, a phase known as the 'asymptotic giant branch', now burning helium in a shell above the remnant core, which consists of carbon and oxygen. Finally, after exhausting shell burning, the Sun will eject its outer layers to form a ghostly 'planetary nebula' around the very compact hot core that is made of the ashes of the nuclear fusion processes. An example is shown in Figure 17, the famous 'Ring Nebula'. At a distance of 2,300 light years, this nebula is easily visible with a modest telescope (a minimum of 10 centimetres (cm) diameter) in the summer constellation of Lyra. These diffuse nebulae can extend as far out as our own enormous Oort cloud. The Ring Nebula has a diameter of about 60,000 AU. By comparison, this is fifteen times larger than the largest of the discs seen in the pre-main sequence object HL Tauri (Figure 15).

The Sun will then traverse across the top of the HR diagram, then move down to below the main sequence (see the track marked in Figure 16). As the nebula expands away from the centre, it leaves behind a 'white dwarf' star that forever radiates remnant heat. The

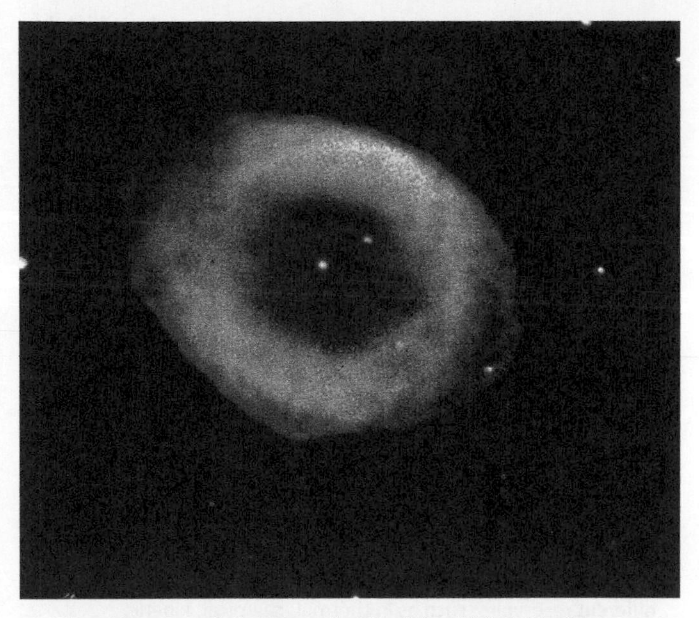

17. The planetary nebula M57, called the 'Ring Nebula', in the constellation of Lyra (the lyre) as it was photographed by the 200 inch Hale Reflector on Mount Palomar, for decades the largest telescope in the world. The Sun will produce a planetary nebula after it has exhausted all its interior nuclear fuel on the main sequence as well as ascending the giant branch twice. The nebula is about 2,300 light years away, it has a diameter of 60,000 AU (compare with the far smaller discs shown in Figure 15).

star that collapsed and led to the Ring Nebula is clearly seen in the middle of Figure 17. While a 10 cm diameter telescope can reveal the Ring, a 30 cm diameter telescope would be needed to glimpse the white dwarf. The central white dwarf has a diameter of 7,000 km, close to that of Earth.

This dwarf is nothing other than the core of the star supported by the quantum mechanical effect of electron degeneracy, and radiating via the slow release of gravitational energy. Remarkably soon after Fermi's work, Ralph Fowler (1889–1944) described the collapse to degeneracy and identified the observed white dwarf stars as the result of the effects of gravity and quantum mechanics. Physically, the slow loss of heat in the white dwarf phase is characteristic of a system that has no access to free energy, and is destined merely to fade away through radiation. The steady increase in entropy marks an observational signature far from the remarkable phenomena we encountered earlier, as the Sun tapped into the free energy of fusion, rotation, convection, and magnetism to produce the unanticipated phenomena we observe, and wonder about, today.

'Free energy'

Free energy is a useful concept in physics and chemistry. It does not refer to the colloquial meaning of getting energy for free from the electrical company! Originally applied to thermal systems, in science it refers to that fraction of energy stored in different reservoirs, such as in thermal, chemical, kinetic, gravitational, or magnetic energy, that is available to do work and be transferred to another form. The concept of free energy was first introduced by Hermann von Helmholtz (1821–94) in 1882. Imagine stretching an elastic band, and twisting it so that it can be used to drive a propeller of a model plane. The energy stored in the twist is freely available to drive the propeller. But the energy associated with the stretching is not. Similarly, although we have vast amounts of energy in our bodies (remembering that $E = mc^2$), none of this 'rest mass' energy is free to use in our everday lives (it would be a very bad thing). We *can* extract chemical bond energy to contract our muscles, but, unlike in the solar core, the nuclear energy stored by the strong nuclear force in our atoms is inaccessible to us.

In another example of the cliché that is the 'circle of life', the loss of the material in the planetary nebula to space amounts to a reseeding of the Galaxy's soil in the stellar nurseries. But this material is now enriched by more elements heavier than iron that result from special nuclear processes occurring on the asymptotic giant branch, the 's-process' elements.

The history of astronomy is often punctuated by particular discoveries and articles. Margaret Burbidge (1919–), Geoffrey Burbidge (1925–2010), William A. Fowler (1911–95), and Fred Hoyle published a paper in 1957 entitled 'Synthesis of the Elements in Stars', later referred to as 'B2HF'. At the time, George Gamow argued that all the elements in the Universe were formed during the 'Big Bang'. But Hoyle argued in 1946 and 1954, and in his B2HF, based upon measured abundance distributions from astronomical objects, that elements heavier than lithium are synthesized during stellar evolution. Of the needed physical processes involved, the capture of free neutrons by specific nuclei was a critical ingredient. Hoyle in 1956 and in B2HF defined two processes: the 'r-process' and 's-process', for rapid (0.01–10 seconds) and slow (100–100,000 years) neutron capture. The time scale for natural (β-) decay of nuclei governed by the weak nuclear force, in particular of isotopes of neon and iron, determines which elements will be produced from the two kinds of processes. The 'r-process' produces elements during the death throes of massive stars (above three times the Sun), during the explosion of supernovae created by the runaway nuclear energy release caused when electron degeneracy pressure can no longer support the overlying weight. Such elements include germanium and bromine. The Sun, during the second ascent of the giant branch, will produce 's-process' elements such as mercury and lead.

The above picture describes the evolution of stars which arrive on the main sequence with masses less than 3-4 times the solar mass, and which exit the giant phase with masses below 1.4 of the Sun's mass. The evolution of more massive stars, such as those which produced the interstellar clouds from which the Sun was born, is different in many details. But it is also an important part of the remarkably successful story of our unfolding understanding of the Universe.

The solar system will undergo radical changes simply as a result of the long-term evolution of the Sun. In terms of life on Earth, there are many other processes, physical, chemical, biological, astronomical, and of course political, that humankind must survive before a significant increase in solar luminosity (10%, say) will drive solar influences to threaten life. Multi-cellular life in principle can survive for about 800 million years before solar driven irreversible changes to the surface chemistry occur. The planets can remain in their orbits as a result of solar changes possibly as late as eight billion years from now, until the inner planets fall into the Sun. At this stage life might be supported briefly on Saturn's largest moon Titan, as it may then support liquid water on the surface.

Chapter 3
Spots and magnetic fields

'On the Probable Detection of a Magnetic Field in Sunspots' is the title of a 1908 paper that has sparked over a century of intense, continuing research. Hale's tentative words are the epitome of a measured, conservative approach to science. In fact, Hale demonstrated that sunspots are magnetic, and we know now that magnetism lies at the heart of what makes the Sun so interesting and important for modern society. Yet, we still do not know why the Sun is obliged to form such strong concentrations as spots at all. How does the Sun reverse its global magnetic fields every twenty-two years? Why must sunspots appear in their characteristic butterfly-wing pattern against time and latitude? Why do polar magnetic fields reverse in fashion synchronized with sunspots? Why must the Sun form a corona? Why must it flare and eject magnetized ionized gas into interplanetary space?

The Sun is a complex non-linear system. It may not be as complex as, say, the human immune system. But, as in biology, our desire to find cause-and-effect is hampered by sheer complexity. Thus, both solar and immune studies are driven not by arguments from first principles, but by *experiment and observation*. Computer models can aid greatly. In the Sun, our understanding of important processes is fairly complete, unlike the vastly more intricate immune system. We can treat the numerical models like 'Real Experiments', and observe and probe them to understand how certain processes might work in the 'Real Sun'. However, the coupling between large and small scales presents a formidable computational challenge for numerical work. The best computational work is treated not as if the calculations were the real Sun, but as experiments in themselves, to be examined, re-run, and understood. But absolutely realistic numerical experiments cannot be performed on more than a small part of the Sun. Just as computational weather forecasting is limited because of the coupling between events on small scales (the 'weather' created by cities) and large scales (the weather over a continent), there are many practical problems that appear intractable for decades to come.

Therefore solar physics remains an observationally-driven subject. In this chapter we will examine what observations tell us about its hidden, inner workings, introducing theory as needed.

Sunspots

In 1843, S. Heinrich Schwabe (1789–1875) found a decade-long period in the annual number of sunspots from his own observations, beginning in 1825. His work is the first report of a cyclic variation of the numbers of spots with time. Six decades later, Annie Maunder (née Russell, 1868–1947) and her husband Edward Maunder examined the spots as a function of solar latitude and time. Their hand-drawn figure, the 'butterfly diagram', is reproduced in Figure 18. The figure is astonishing to physicists, who would ask

> *Out of the chaos of a convecting star, why would the Sun choose to do this?*

The story behind the figure is itself interesting. According to Thomas J. Bogdan (1957–), Annie Maunder had reported:

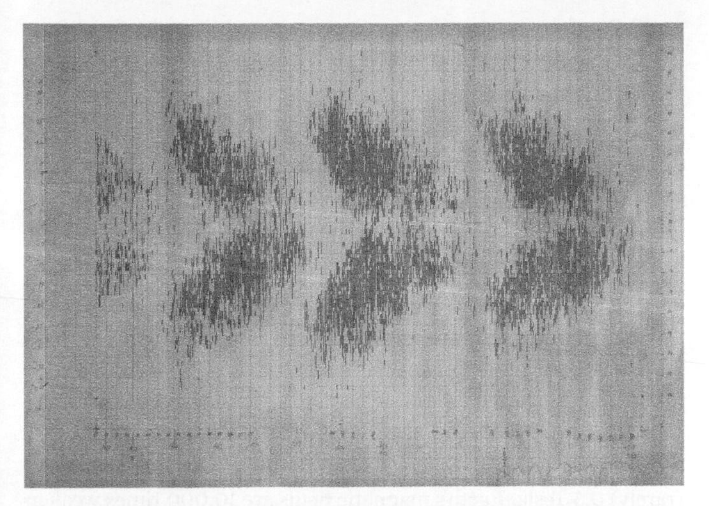

18. A hand-drawn plot of the appearance of sunspots on the solar disc, by Annie and Walter Maunder in 1904. Time runs from left to right over a period of about thirty years, and latitude runs from south pole to north pole of the Sun. It currently decorates the wall of the HAO of the National Center for Atmospheric Research in Boulder, CO, USA.

> We made this diagram in a week of evenings, one dictating and the other ruling these little lines. We had to do it in a hurry because we wanted to get it before the [Royal Astronomical] Society at the same meeting as the other sunspot observers, whose views we knew to be heretical. As it turned out the diagram wiped [the other observers'] papers clean off the slate.

On 21 May 1940, Annie mailed the drawing from London to a friend in the USA, to save it from possible destruction during the Blitz. The picture of typical sunspot behaviour became essentially complete by 1920: Hale and colleagues in California had measurements of magnetic and statistical properties of sunspots, uncovering several statistical 'laws':

- Spots always appear in pairs containing opposite polarities, a preceding spot (p) and follower (f), lying roughly along the solar east–west line;
- 'Hale's law' states that the p-spot of a pair has opposite poles in opposite hemispheres;
- 'Joy's law' (after Alfred Joy; 1882–1973) states that the pairs emerge with the p-spot closer to the solar equator, that is, the magnetic bipole is tilted from the east–west line;
- 'Spörer's law' (after Gustav Spörer; 1822–1895) states that new sunspots emerge first at high latitudes ($\pm 30°$) and progressively move towards the equator over several years.

Most spots are between 0.1 and 4 per cent of the Sun's diameter across. They have a field strength between 0.2 and (rarely) 0.5 Tesla. Earth's magnetic fields are 10,000 times weaker, but strong household magnets are up to 1 Tesla. Sunspot field strengths are limited in strength by the need to balance surrounding pressures with the magnetic and plasma pressure within the sunspot.

Figure 19 collects modern estimates of counts of sunspots over recorded history. The pitfalls in such studies can be significant. These modern compilations are anything but straightforward. Complications arise because different observers report different numbers of spots, different telescopes have different characteristics, and the observability of spots depends not only on weather but also on 'seeing', the blurring of images by convective bubbles of air in Earth's atmosphere. But having undergone many re-analyses, researchers believe that modern plots are fairly and accurately portraying sunspot numbers over four centuries. The dominant signal is the eleven-year period of the waxing and waning of sunspots. We readily accept these regular oscillations today, not because we understand why the Sun must do this, but because they have become familiar. Below I will collect arguments showing that rotation and convection are two

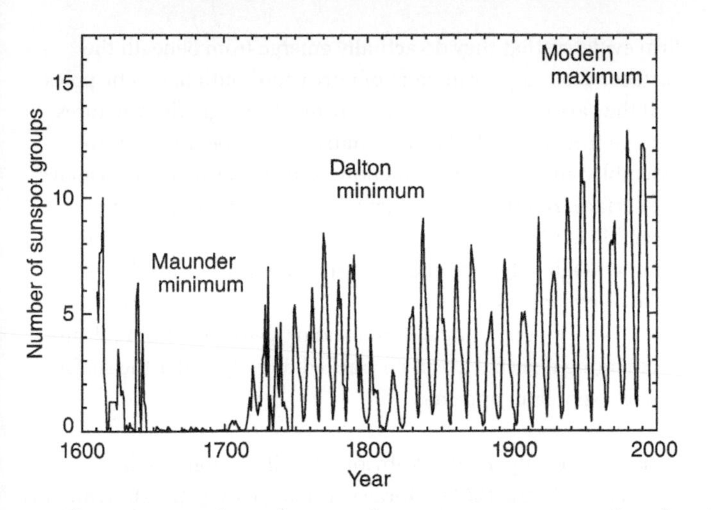

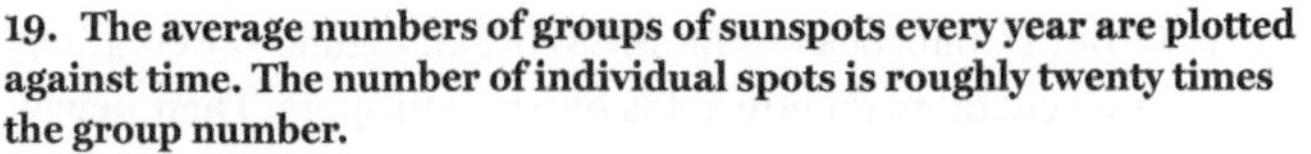
19. The average numbers of groups of sunspots every year are plotted against time. The number of individual spots is roughly twenty times the group number.

necessary ingredients needed for the Sun to exhibit such behaviour.

The figure also reveals strong modulations of this oscillatory cycle, including the prominent 'Maunder Minimum' and later 'Dalton Minimum'. As defined by Jack Eddy (1931–2009), the Maunder Minimum extended from *c.*1643 to 1715. Outside of these dips, the orderly coming and going of sunspot numbers is remarkable. The Sun produces sunspots and organizes them into robust, repeatable global-scale patterns. Even during the Maunder and Dalton minima, natural records on Earth (tree-rings and ice-cores discussed below) reveal that the prominent eleven-year period in sunspot numbers persisted, but with relatively smaller solar signals.

When viewing the birth of sunspots, they seem to emerge, essentially complete, from within the solar interior. But is there firm evidence that they do actually emerge from beneath the surface? They appear in pairs of north and south magnetic poles as if they are places where sea-serpent-like magnetic structures emerge from beneath. The alternatives could be that they are assembled in the surface somehow, or that they come from above the surface. But there is no sign of surface flows that suggest assembly in place (in fact the opposite is generally seen), and there is not enough magnetic energy in the overlying atmosphere to build a sunspot from above. Direct evidence of emergence has recently been seen using helioseismology, where blurry indicators of emerging objects have been tracked directly under the surface where a spot later appears.

Sunspots are only the most obvious manifestation of solar magnetism. In the 1950s, Horace W. Babcock (1912–2003) and his father Harold D. Babcock (1882–1968) began regular observations of the more general magnetic fields outside sunspots. Their newly developed system using electronic detectors was able to record weaker magnetic signatures using the Zeeman Effect, by observing

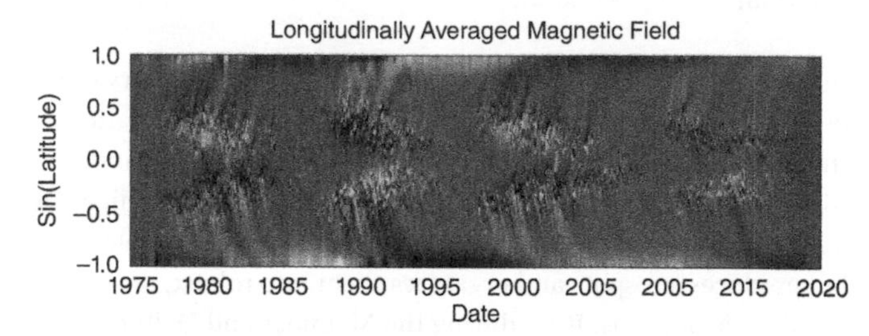

20. Magnetic fields recorded and stacked in time reveal the sunspot butterfly diagram, but in the context of the more general solar magnetic field. The *y*-axis shows sin latitude, 0.5 corresponds to 30 degrees latitude, 1.0 to the north pole. Sunspot magnetic fields are seen as grain-like spots of positive and negative polarity, and they are superimposed on a smoother background of near-zero magnetic fields (grey colours). The fields at the poles change sign every eleven years (white–black) so that the total cycle takes twenty-two years on average. A white point is +10, black -10 Weber m^{-2}.

polarized light. As discovered by Zeeman, the presence of magnetic fields in a gas changes both the wavelengths (as measured in 1908 by Hale) and polarization of the light in spectral lines. Just as we can use polarized glasses to remove reflected glints from streams to see fish better, we can use polarized spectra to see far weaker or spatially diluted magnetic fields (field concentrated into areas smaller than a camera pixel). A picture of the Sun made in this fashion is called a 'magnetogram'. A particularly revealing form of magnetogram is shown in Figure 20, in which cuts along the north–south axis of the Sun are measured every day, and then stacked in time. The plots reveal the evolution of more than the sunspot fields. The impression from such figures is that of a time sequence:

- systematic spot emergence between -30 and 30 degrees latitude, within the 'wings of the butterfly' (Figure 18);
- leading and following spots are of opposite signs in each hemisphere (Hale's polarity law);
- steady motion of diffuse magnetic fields from spots towards the poles, over a year or so;
- 'cancellation', or reversal of sign of pre-existing polar fields, following the poleward motion of follower sunspots emerging systematically closer to the poles ('Joy's law');
- after eleven years, the entire pattern is repeated with polarities now all reversed.

These observational records (Figures 18 and 20) represent the prime observational hurdle that any theory of the evolution of solar magnetism must overcome. A stripped-down summary of these essential properties is shown in Figure 21. It highlights 'Joy's law', showing the poleward tilt of follower spots to leader spots, which are connected by the dashed lines. Note that the tilt is always *slight*, a few degrees from a line of constant latitude, and that the sunspots closest to the pole in each hemisphere are opposite in polarity to the polar fields.

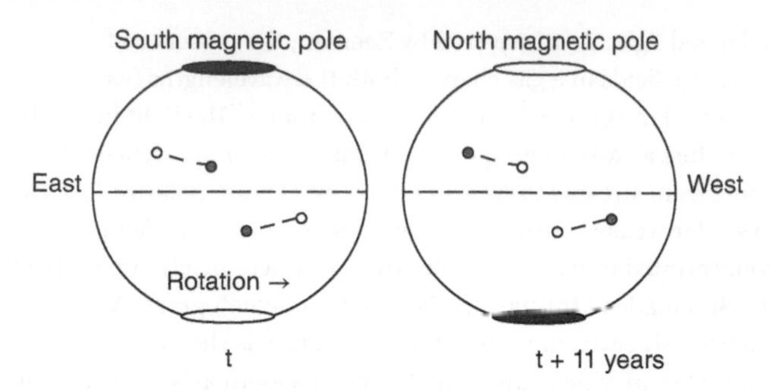

21. A condensed version of the dominant behaviour of observed solar surface magnetic fields during two opposite 'phases' of the solar cycle. The two phases shown might correspond to around 1988 and 1999 in Figure 20. The white shading has the opposite magnetic polarity to the dark shading. Sunspots are shown as small circles.

The Sun among the stars

The orderly behaviour represented by Figures 18 and 21 is so extraordinary that astronomers have repeatedly asked, how does this come about? Is this behaviour typical of other stars? Obviously, many years of observations were needed to obtain the results of Schwabe, the Maunders, Hale, and many others. Data spanning such long periods of time are not available for stars. But, thanks to some forward-looking scientists, we do have over five decades of related stellar data of vital importance.

Every few years, an astronomer seems to ask the question, paraphrased by Bengt Gustafsson (1943–):

> *Is the Sun an oddball?*

The question is usually prompted by the knowledge that the only life is on its third planet. Is there something special about our star? Since 1988, we have detected planets around other stars and the numbers have exploded with space experiments such as *KEPLER* and *TESS* unearthing several thousand more. Based upon measurements of many other stars, the Sun itself appears to be unremarkable. Maybe. Let us look at how the Sun stacks up against other stars. Phrased differently, let us re-run the 'solar experiment' using an ensemble of stars, to reassure ourselves that we do not have to deal with a genuine oddball.

We cannot resolve any detail on the discs of Sun-like stars. The nearest Sun-like star, α Centauri A, is about 20 per cent larger than the Sun, and four light years distant. The Sun, only eight light minutes away, appears about 200,000 times bigger. To see the largest sunspot ever seen, placed on α Cen A, we would need a telescope (in space) with a diameter of about 700 metres (for comparison the Hubble Space Telescope has a 2 metre diameter mirror). Worse still, we cannot observe directly the magnetism in any spot on α Cen A, because of the absence of magnetic monopoles. If we were to sum up the signals in Figure 20 in latitude, to try to derive a total magnetic field as a function of time for the Sun, we would find an answer far closer to zero than the variations. As Hale found, the spots come in pairs, and the p and f spot pair almost cancels out each other's Zeeman signal. Only in special cases are we able to track magnetism through the Zeeman Effect in a star like the Sun. One case is that we might detect a star that we fortuitously see with its poles pointing in our direction. Looking at the Sun in this fashion, we might hope to detect the north or south patches at the polar caps (Figure 20).

But we *can* observe effects of magnetic fields associated with spots that do *not* cancel out. It might be thought that the visible Sun gets dimmer as a dark spot passes across the disc, as light is perhaps blocked by the suppression of convection that causes the spot to be cooler than the surroundings. This is sometimes, but not often, true. In fact, the opposite is more often the case, because neighbouring regions surrounding spots become brighter (see Figure 22). If we look at variations in the brightness of the entire Sun at all wavelengths, the changes are typically 0.04 per cent, with occasional excursions of 0.1 per cent. These tiny fluctuations

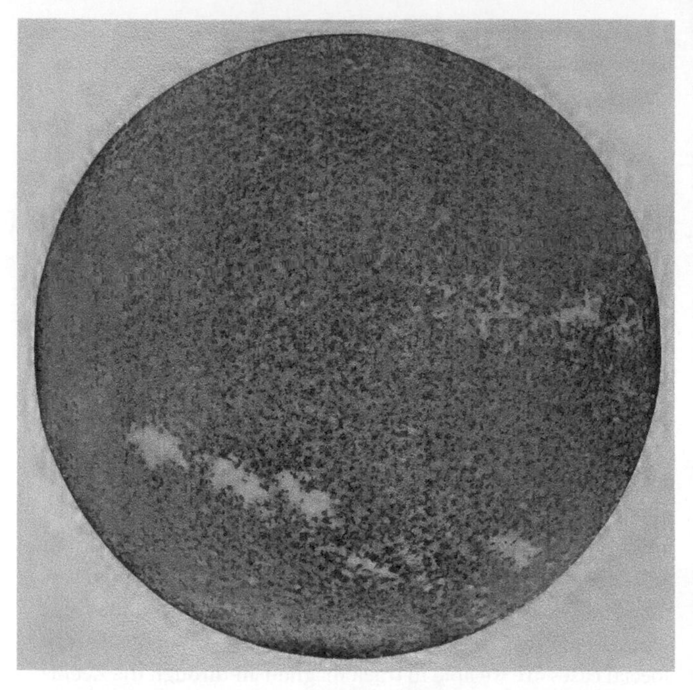

22. A 1903 image near the centre of Fraunhofer's H-line of singly ionized calcium, formed in the solar chromosphere. It was taken by Hale and Ellerman in 1903, using the Rumford spectroheliograph, Hale's second generation instrument. Note the bright clumps that are associated with sunspots; these are the first photographic records of plages.

are measured by modern instruments in space. Such variations are very difficult or impossible to see from the ground, as our atmosphere introduces spurious variations in transparency, scattering, and clouds. Yet such variations, surprisingly, can be observed in stars. How so? Well, stars are point sources and we can compare Sun-like stars with other, more stable groups of stars.

In Chapter 1, we saw that magnetism pokes through the visible surface and into less dense layers. While we do not yet understand in detail why, magnetic energy above the visible surface contains free energy that is turned into heat, and then UV radiation, to form the chromosphere and corona. The free energy is a positive quantity, so it leads to a brightening when converted to heat, for instance in the chromosphere. The detailed association of bright Ca^+ emission with surface magnetic fields was established in 1959 by Leighton, using sensitive measurements of magnetic fields. Hale and colleagues found that Ca^+ emission surrounds dark spots (Figure 22), superposed on an ever-present network of Ca^+ emission over the rest of the Sun.

The bright sunspot regions are called plages (Fr. *beaches*). The passage of plages across the Sun as it rotates could readily be measured by an alien civilization elsewhere in our Galaxy, merely as changes of brightness, revealing a clear magnetic signature. In turn, we can look for Ca^+ variations on other stars. Other lines are also enhanced in plages, but the Ca^+ lines are the easiest to measure on other stars. Astronomer Olin Wilson (1909–94) began a survey of Ca^+ emission of a variety of bright stars, at Mount Wilson Observatory in 1966 (the names are a coincidence). Figure 23 compares solar behaviour with some typical stars, as seen in Ca^+ variations.

Seen in this fashion, the Sun's eleven-year cycle of spot behaviour turns out to be a little unusual. In his 2017 PhD thesis, Ricky Egeland (1980–) collected and re-calibrated broad and diverse sets of data from several projects, mainly from the 'Mount Wilson' survey and the 'SSS' project run at Lowell Observatory by Jeffrey Hall (1964–) and Wesley Lockwood (1941–), delivering data since 1994. Egeland discovered that the Sun has the highest quality cyclic behaviour of any star, as measured through the simplicity and repeatability of light curves, and that its cycles are characteristic of cooler, K-type stars on the main sequence. Such stars are of a lower mass, and they have a deeper convection zone in the interior, compared with the Sun.

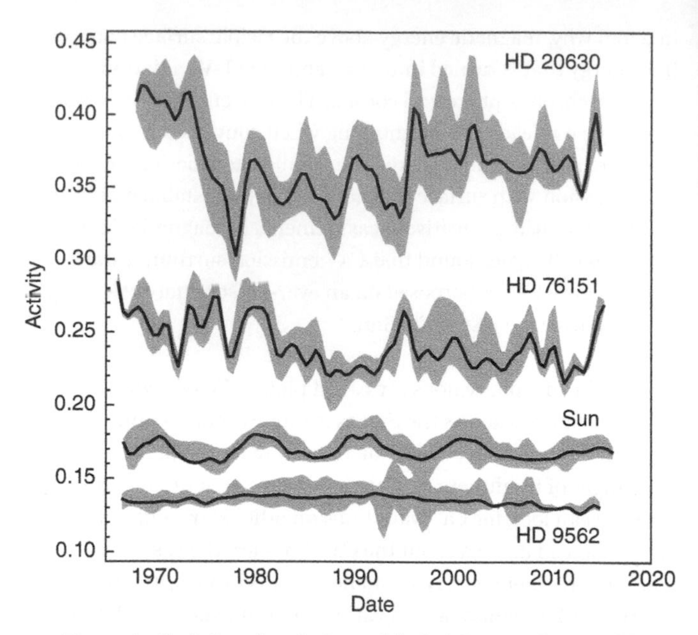

23. Magnetically-induced variations of the brightness of spectral lines of Ca⁺ are plotted as a function of time, for the Sun and representative stars. These variations are caused by the coming and going of star spots. The *y*-axis measures the brightness of Ca⁺ emission lines, a 'proxy' for surface magnetism (Figure 22), relative to the brightness of each of the stars.

Skumanich's 1972 results showed that the average brightness of Ca^+ lines appears to vary as the inverse of the square root of a star's age. The implications for magnetism were clear: as the Sun evolves, it loses plages associated with sunspot magnetic fields. Decaying quickly at first, they slowly tail off with age on the main sequence. The Sun must have been far more magnetically active in the early solar system.

A decade later, in seminal work led by Robert Noyes (1934–) in 1984, a deeper connection between magnetism and rotation was uncovered. Combining models of interior convection with measured brightness of Ca^+ and rotation rates for a larger number of stars, Noyes discovered that the Ca^+ emission followed best a particular combination of physical parameters. The relevant parameter on the x-axis of Figure 24 is

$$x = P/\tau,$$

where P is the period of rotation of the star, and τ is the time taken for one convective element to 'turn over' (roughly a lifetime for one element). The authors found that, of many other combinations, this combination yielded the tightest correlation. Now τ is not an observed quantity, but one taken from theory and numerical models. This is an example of a common challenge facing astronomers: we are often obliged to augment data with other information, such as from theoretical models, in order to make progress.

The significance of this plot is many-fold. First, there appears to be a parameter, P/τ, which correlates well with the Ca^+ emission, strongly suggesting a direct link between magnetism and P/τ. Second, this spans almost an order of magnitude in both the brightness of Ca^+ and in this parameter. Third, it smoothly joins the groups of 'old' and 'new' stars. In other plots these were separated. Last, it ties together concepts from dynamo theory which we will encounter shortly. The ratio P/τ is called the 'Rossby number', *Ro*, after meteorologist Carl-Gustaf Rossby (1898–1957). It measures the influence of rotation on blobs of convecting fluid.

Records of solar magnetism

Spots on the Sun became a target of serious attention shortly after the invention of the telescope. The first written record (1608) of a refracting telescope is due to Dutchman Hans Lippershey (1570–1619). By projecting the Sun's image on to a flat surface, several observers, including Galileo, Christoff Scheiner (1573 or

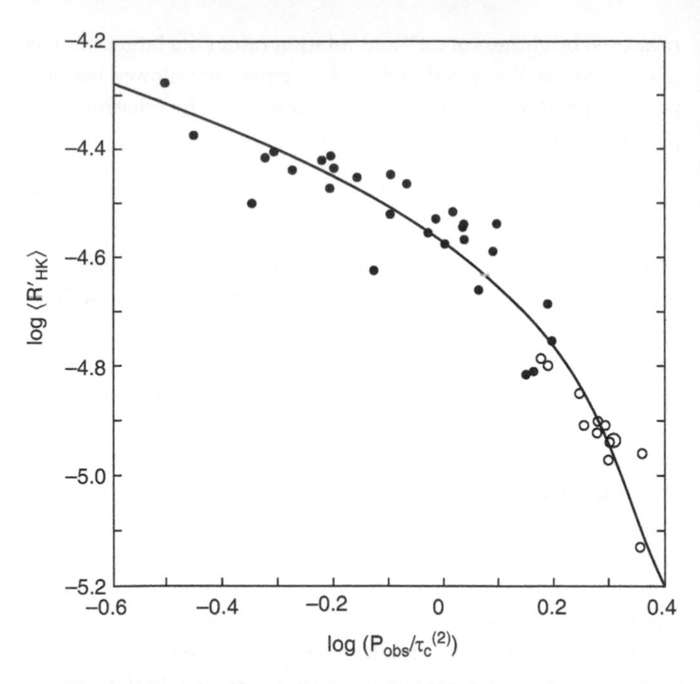

24. The fraction of stellar radiation emitted by Ca^+ ions for a sample of forty-one stars is plotted as a function of the ratio of the period of rotation of a star to the 'convective turnover time'. The Sun is shown as the large open circle. Filled circles are young stars, open are old stars. In this context the Sun is 'old'.

1575–1650), David Fabricius (1564–1617), and his son Johannes (1587–1615) were observing and tracking sunspots. The 17th-century work has been shown to be of high quality, permitting modern researchers to quantify sunspot records back to Galileo. At the opening of a readable and influential article, in 1976 Eddy concluded that 'The reign of Louis XIV [1643–1715] appears to have been a period of real anomaly in the [magnetic] behaviour of the Sun'.

Gustav Spörer was the first to note a prolonged period of low sunspot activity from 1645 to 1715, the Maunder Minimum. In collecting data of sunspots and aurorae in the 1970s, Eddy noted that the modern eleven-year cyclic variation of spots essentially disappeared, along with a dearth of aurorae. Eddy's work has been subject to intense scrutiny, but his main conclusions have survived more-or-less intact. When present, the spots during the Maunder Minimum appeared almost always near the equator, in the southern solar hemisphere. A modern estimate of sunspot counts is shown in Figure 19. The Sun also hiccuped between 1790 and 1830, a period named the 'Dalton Minimum' after John Dalton.

Written records of solar magnetism are limited to 1608 onwards. But history is often written by nature. There are two 'cosmogenic' isotopes which provide indirect records of solar magnetism. The mechanisms behind these records are interesting. Cosmic rays generated in very energetic events from across the Universe enter our solar system at random, from all directions. The origin of these very high energy atomic nuclei is still being researched. But their trajectories within the solar system are influenced by the entire solar magnetic field throughout its sphere of influence—the 'heliosphere'. The solar-modulated cosmic rays enter the atmosphere and cause 'air showers', chains of nuclear reactions. Such reactions lead to special isotopes which rain out of the air to record, crudely, the state of the heliospheric magnetic field. Trees and ice cores record the different amounts of these isotopes every year, giving us a record of the Sun's modulation of cosmic rays going back some 10,000 or more years. The physical links are as follows:

$$\begin{aligned}\text{cosmic rays} &\rightarrow \text{modulated by heliospheric magnetism}\\ &\rightarrow \text{atmospheric nuclear reactions}\\ &\rightarrow \text{precipitation}\\ &\rightarrow \text{recorded in tree rings } (^{14}\text{C}) \text{ and ice cores } (^{10}\text{Be})\end{aligned}$$

Direct measurements of atmospheric nuclear reactions have been made since about 1950, using 'neutron monitors'. The tree ring and ice-core records can thus be 'calibrated' using such measurements, and the sunspot record extrapolated back in time. The records are imperfect for several reasons, but two isotopes, ^{14}C (half-life 5,730 years) and ^{10}Be (1.4 million years) are uniquely formed in this fashion, and so problems associated with local climate conditions are minimized. The largest uncertainties arise from rates of variable precipitation of isotopes out of the atmosphere. Several groups worldwide have meticulously analysed such data. Figure 25 shows a reconstruction of the heliospheric magnetic field near Earth over the holocene, the 10,000 years following the last extensive glaciation.

So, as far as we are able to measure, the Sun's magnetic field and its variations generally fall into line with other stars in our Galactic

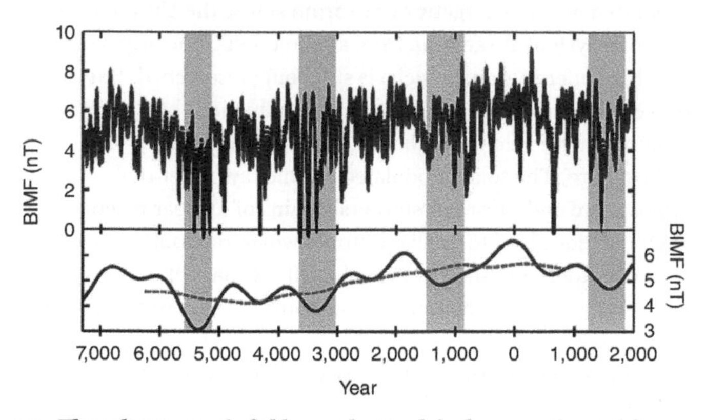

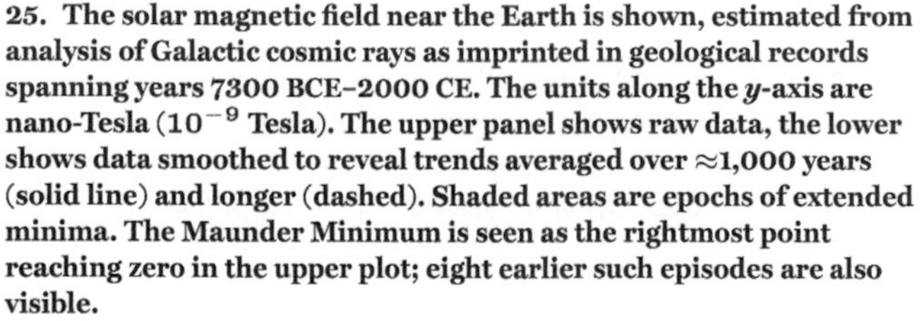

25. The solar magnetic field near the Earth is shown, estimated from analysis of Galactic cosmic rays as imprinted in geological records spanning years 7300 BCE–2000 CE. The units along the *y*-axis are nano-Tesla (10^{-9} Tesla). The upper panel shows raw data, the lower shows data smoothed to reveal trends averaged over ≈1,000 years (solid line) and longer (dashed). Shaded areas are epochs of extended minima. The Maunder Minimum is seen as the rightmost point reaching zero in the upper plot; eight earlier such episodes are also visible.

neighbourhood. As reported by Egeland, its eleven-year cycles during the recent modern maximum (Figure 20) are the most consistent of all stars measured so far. However, this has not always been the case. The Sun experienced the Maunder and Dalton minima, and the modern maximum during four centuries of documented history. Radionuclide records in ice and tree rings show that the sunspot history of the past 400 years is representative of the past 9,000 years.

Dynamos

A dynamo is a well-known device for generating an electric field, a voltage V volts, by moving wires across the magnetic field of a permanent magnet. The charges in the wires experience an electric field owing to Faraday's law of electromagnetic induction.

Electrons are only weakly bound to the metal ions. Being light they readily move, and the motion relative to the metallic ions constitutes the electric current. If the dynamo circuit is open (switched off) but the wires still move, the electrons quickly move until they make their own electric field that opposes and balances the voltage V. But when a circuit is connected through a resistor of resistance R, such as the filament in a bicycle light, currents can flow and the electrons move and collide with the atoms in the resistor. The KE of the moving wires (the armature) is converted to electrical energy and then dissipated in the resistor to emerge as heat and light.

In an electrically conducting fluid, a dynamo is a little different, but the principle is the same. Both ions and electrons comprising the fluid are free to move in response to magnetic and other forces (gravity, viscosity, pressure gradients). Each blob can sustain a magnetic but not an electrical field within it. As before energy is converted from mechanical motions into electromagnetic energy. But unlike the wire, KE is overwhelmingly converted into *magnetic* energy.

Faraday's law of electromagnetic induction

In 1831 Faraday discovered that a changing magnetic field, made by moving a magnet relative to a wire, induced an electric current in the wire. Now, currents flow only upon application of a voltage or electromotive force, and Faraday's law of electromagnetic induction states that 'the electromotive force around a closed circuit is equal to the negative of the time rate of change of the magnetic flux enclosed by the circuit'. (Flux is the product of the magnetic field times the area.) The law is encoded directly into Maxwell's successful unified theory of electromagnetism.

The law is at the heart of many devices such as transformers. Yet it is intimately tied to Einstein's Special Theory of Relativity. His famous 1905 paper begins noting that Faraday's law applies only to the *relative* motion between magnet and wire. It did not matter whether wire or magnet were moved. Upon this fact and the observation that the speed of light is the same as measured by all observers, no matter their speeds, Einstein built his Special Theory. Thus the humble transformer is a direct manifestation of Special Relativity!

In a large, highly conducting fluid like the Sun, this effect is central to the fluid's dynamics. *Any* motion of fluid across magnetic fields lines (or vice versa) generates an electric field. When there is no relative motion between fluid and magnetic field, as we have seen, any electric fields are shorted out by the mobile electrons. But this new 'induced' electric field persists. Each charge 'sees' an induced electric field only if it moves across and relative to magnetic field lines. The induced 'electric' force is none other than the magnetic Lorentz force, written earlier as $q[\mathbf{vB}]$. In other words the charges appear to experience electric and magnetic forces depending on their velocity relative to the magnetic field. In fact they experience just one force, that of electromagnetism, which we experience daily as electricity and magnetism.

This is not (quite) the end of this comparison between a dynamo with wires and one in a fluid. In the bicycle-light case, energy flows only one way, in the fluid case, it is a two-way exchange, as follows:

Bicycle-light dynamo

muscular chemical energy $\Rightarrow$ drives bike & armature wires
wires cross bar magnet fields $\Rightarrow$ induces electric field
electric field $\Rightarrow$ drives current
current dissipated $\Rightarrow$ heat & light

But in fluids there is a complication, and it is a big one. The electrons and ions experience the Lorentz force as they move across the magnetic field. (The same happens in the wire of the armature in the bicycle dynamo, but the electrons must flow along the rigid lattice that comprises the solid wire.) Since energy is conserved, the Lorentz force must ultimately limit the operation of a dynamo, otherwise we would have a perpetual motion machine. This MHD dynamo works as follows:

MHD solar dynamo

nuclear & rotational energy $\Rightarrow$ **fluid motion**
fluid motion $\Rightarrow$ magnetic fields
magnetic fields $\Rightarrow$ Lorentz force on fluid
Lorentz force $\Rightarrow$ alters **fluid motion**

Physicists call this situation 'dynamic', in that the motions and magnetic fields evolve *together*, and unless the magnetism is everywhere very weak (i.e. the last two steps are negligible), the evolving fluid velocities, densities, temperatures, and magnetic fields must generally be expected to show non-linear, complex behaviour. Sometimes the Lorentz force can be neglected. In this 'kinematic' case, just as for the bicycle light where the faster you go, the more light is generated, the more energetic the prescribed motions, the larger the magnetic fields generated. But the Lorentz force cannot be neglected without great care. The strength of sunspots fields is limited by the ability of the surrounding plasma to contain the sunspot magnetic stresses. The dynamical approach is the more complete theory.

Solar magneto-hydrodynamic dynamos

A MHD dynamo can be defined as:

> a conducting fluid system in which magnetism is sustained at a finite level as time proceeds, against the effects of resistive dissipation.

This is essentially a statement about just the magnetic energy, without regard to constraints from, for example, topology. Clearly the observed Sun satisfies this definition. But the observed ordered magnetic fields of the global Sun implies that the underlying dynamo must be rather special.

Magneto-hydrodynamics and topology

The simplest description of plasma is as a *magnetized fluid*, a fluid that can sustain large-scale magnetic fields but not electric fields. A highly-conducting magnetized fluid has remarkable *topological* properties. Two geometric objects are topologically identical if one can be continuously deformed into the other. A doughnut has the same topology as a chicken. The relevance of topology to the Sun arises from a theorem first derived by Nobel Laureate Hannes Alfvén (1908–95) in 1942:

In a fluid with infinite electrical conductivity, magnetic fields are frozen into the fluid and must move along with it.

The theorem arises from Faraday's law of electromagnetic induction and the Lorentz force.

In trying to drag magnetic lines of force across a fluid blob, electrons and ions try to move in opposite directions because of Faraday's induced electric field setting up a current. But no current can flow because the electrons move immediately to short out the field, unless other forces alter this motion significantly. One such possibility is the effect of collisions between the particles which lead to a finite conductivity. But in the absence of such forces (a situation we call 'ideal' MHD) there can therefore be no motion across magnetic field lines.

The situation leads to the key concept of a tube of magnetic flux. Imagine a blob of fluid threaded by magnetic lines of force. Follow the lines in space far enough that they return to the blob at the other side (no monopoles). Then the *'flux tube' is the torus (doughnut) encompassed by the outer surface of all these field lines.* Each magnetic tube is defined for all time by the fluid within it. We could dye the blob, and the dye would forever trace the same flux tube. We see that

the topology of magnetic fields in an infinitely conducting fluid is fixed, for all time.

This corollary to Alfvén's theorem is a central concept in solar MHD, where the conductivity is high, the fluid almost ideal, and physical scales are large.

Solar/stellar research indicates that the solar dynamo obtains *energy* from rotation and convection, because observations (Figure 24) point to the importance of a particular combination of rotation period P and convective turnover time τ, through the Rossby number, $Ro = P/\tau$. But how might this work? An isolated, solidly rotating conducting body can only change its magnetic field through resistive decay (non-ideal effects). For a solid body with otherwise solar properties, the decay would take one hundred million years or so. So in seeking a suitable solar dynamo, we must look not to solid rotation, but to fluid motions within the body *related to* rotation. Joseph Larmor (1857–1942) proposed a conducting fluid model for the Sun's magnetic field in 1919. His proposals were generally accepted until Cowling in 1933 derived an important 'anti-dynamo' theorem, contradicting Larmor's earlier ideas:

> *it is impossible that an axially-symmetric magnetic field shall be self-maintained.*

Most stars are, on large scales, almost spherically symmetric, and very closely symmetric around their rotation axis. The trick was then to identify suitable symmetry-breaking processes capable of sustaining a global dynamo.

In 1955, Eugene Parker proposed an MHD dynamo capable of describing the solar behaviour based upon just two ingredients: rotation and convection. The solar surface rotates considerably faster at the equator than the pole. We will see that this behaviour extends throughout the convection zone, so that different regions of the solar *interior* also rotate differently. These flows drag the magnetic field along with them owing to Alfvén's theorem (breaking axisymmetry, and avoiding Cowling's objection). The magnetic tubes associated with a speedy flow will become stretched along with the fluid as it moves around the rotation axis, to catch up with others rotating more slowly. In time, more and more magnetic field is wrapped around the axis (see the top right panel of Figure 26). Magnetic energy is increased by the work done against the Lorentz force. But the number of field lines wrapped around the Sun's rotation axis has increased as well. This is the definition of an increased magnetic field strength. This picture describes how a differentially rotating star can generate strong magnetic fields within a volume in the shape of a torus around its rotation axis. The effect is called the Ω-effect and it is believed to be one essential ingredient of stellar MHD dynamos.

The torus-like fields are of a different nature to those polar fields we see traced out by plasma in eclipse images, which have components not around, but along the rotation axis. These solar 'poloidal' lines of force are visible most readily as the bright rays in Figure 10, near the north and south solar poles (top and bottom of the figure respectively). The roots of these fields are seen at latitudes of $\pm 90°$ in Figure 20. These two kinds of fields, *toroidal* and *poloidal*, can be used to decompose the total, global magnetic field of a sphere. The Earth has a (mostly) poloidal magnetic field—a south pole (near the Earth's north geographic pole) to which the north pole of our magnets point, and vice versa. The magnetic lines of force look like those of a single bar magnet (see the top left panel of Figure 26), joining the north and south magnetic poles, tracing lines of constant magnetic longitude.

In these terms, the Ω-effect converts weak poloidal magnetic fields into strong toroidal fields. Sunspots have surface structure that is essentially that expected if strong toroidal fields were to rise and pierce the solar surface: they are oriented mostly along the east–west direction, and they contain the strongest observed magnetic fields on the Sun. Cowling's theorem can be considered as it applies to toroidal and poloidal fields. Each of these components can be imagined as a flux tube or piece of wire forming a loop. Enforcement of axisymmetry means that each kind of field lives forever in its own plane. In consequence, there can be no way to deform the one kind of field into another without breaking axisymmetry.

So, the critical observational data represented in Figures 18, 19, 20, and 21 have led us to reduce the 'global dynamo problem' to one in which the sequence

$$poloidal \Rightarrow toroidal \Rightarrow poloidal \Rightarrow toroidal \ldots$$

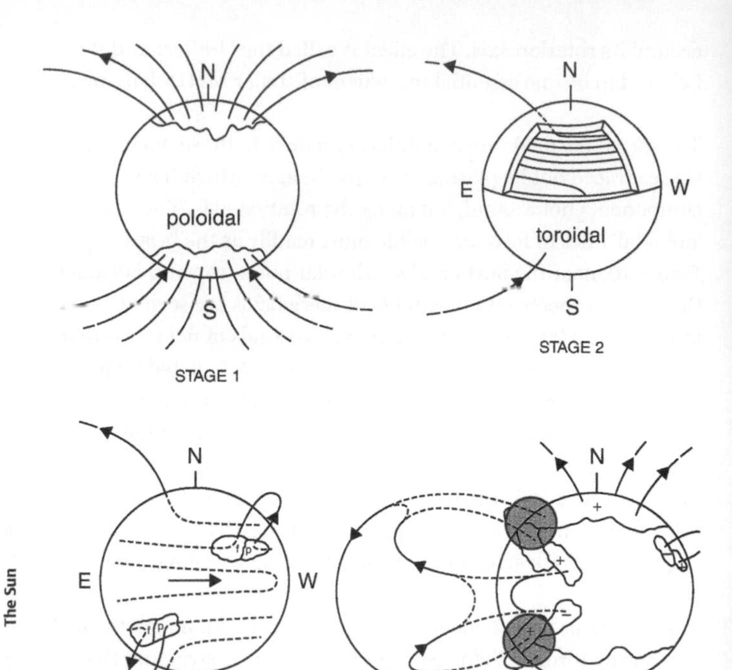

26. The upper panels show the poloidal and toroidal components of magnetic fields. 'Stage 1' shows only poloidal fields (bar magnet). Stage 2 shows their conversion into toroidal fields via the 'Ω-effect', producing field lines wrapped around the rotation axis. Stage 3 shows tilted sunspot magnetic fields emerging through the surface, arising from the toroidal fields in the interior (dashed lines). In Stage 4, relaxation via diffusion and/or reconnection changes the topology of the magnetic fields at and above the surface, leading eventually to the reversal of the polarity of pre-existing polar fields. The shaded circles highlight fields from tilted sunspots that will move polewards to reverse the polarity of the poles.

represents what we should seek as the 'bare bones' of a working dynamo model. The one full cycle represented by the above sequence takes twenty-two years.

Parker then showed that rotation plays another key role in quite a different manner. While the significance of Ro was not appreciated at the time for stars, the importance of it for dynamics of Earth's atmosphere was well-known. Ro measures the ratio of inertia—the tendency for objects to continue in a straight line (Newton's First Law)—to an apparent force that arises when we are moving within a rotating structure. This *pseudo*-force—the Coriolis force—permits us to apply Newton's laws in the simpler, more convenient non-rotating frame, simply by adding a new term to the equations. It is strong when $Ro < 1$, and weak otherwise. If you have tried to walk on a rapidly-rotating roundabout (P is small, a few seconds say), you will know this disorienting force. When the roundabout rotates slowly (P of a minute, say), then we feel little effect. This explains why people feel no such force on the surface of the rotating Earth, and why the ancients considered that the Earth is fixed and the Universe rotated around us.

The particular 'handed-ness' (clockwise or otherwise) of the Coriolis force is a feature of pseudo-forces. We have seen that the Lorentz force—also a pseudo-force—required a choice of direction, mentioned as the 'right-hand rule'. These forces therefore *break rotational symmetry*, a requirement for the operation of any large-scale stellar dynamo.

Parker specifically considered the Coriolis force acting on convection. We can usefully imagine ourselves rising with convecting fluid. We are pushed by pressure differences, pulled by gravity, but because we are rotating we are subject to the Coriolis force. As we rise, inflows fill the void left behind. Some of the motions are horizontal, and the Coriolis force imparts on them a twist around the local vertical direction. The same thing happens as the convection 'runs out of steam' at the end of its trajectory.

This kind of twisting is familiar during hot weather. A large thunderstorm develops because overheated ground warms the air above and leads to convection. Air moves horizontally to replace the rising air, and because the Earth rotates, it experiences the Coriolis force. On very large scales, such phenomena can form a tropical cyclone, moving anti-clockwise in the northern hemisphere and vice-versa in the southern hemisphere. In the Sun, magnetic fields frozen to the fluid experience the same twisting motions. If the toroidal field is dragged upwards by convection, we have a recipe perhaps for cooking up sunspots. The Ω-effect can generate very strong toroidal fields. The Coriolis force that drives Parker's proposed 'cyclones' can twist rising toroidal fields in opposite directions in each hemisphere.

Parker's cyclonic motions in principle achieve two essential parts of a workable solar dynamo: they introduce tilt that might account for Joy's Law in both hemispheres, and this tilt is nothing other than conversion (rotation) of an east–west toroidal field into a north–south component, that is, a poloidal component. So here we might have a way to make the solar cycle work.

The twist or 'helicity' from cyclones was later encoded into mean-field computer models through a parameter called α, which measures how much a flux tube containing some helical magnetic field wraps around itself, like the paint on a barber's pole, per unit length. A 'slinky' has a high helicity, a sailor's rope a modest helicity. Parker's cyclonic motions are equivalent to this 'α-effect'. The global dynamos based upon the two effects we will call '$\alpha - \Omega$' dynamos.

As outlined so far, with the operation of both Ω- and α-effects, we have a recipe for generation of magnetic fields *ad infinitum*. Worse still, given Alfvén's theorem, the topology is preserved on all but the smallest scales, and there may develop a conflict between force balance and topology. These conceptual difficulties suggest a picture that may resolve how the Sun might relax to permit a

Changing magnetic topology

Breaking the frozen-field condition to allow changes of magnetic topology is *required* for dynamo action. *Magnetic reconnection*, observed in laboratory and space plasmas, can enable this to occur on dynamical time scales, much faster than bulk diffusion scales. Figure 27 shows magnetic fields in an 'X-point' configuration that are anti-parallel. As demonstrated in the 1950s and 1960s by several scientists, when brought into close contact, for example by convective motions, they become susceptible to reconnection shown in the central and right panels. From the work of Biot and Savart, the central configuration must carry an electric current in a sheet that flows into the picture, extending along the x-axis. Local resistive dissipation of this current allows magnetic flux and fluid to diffuse through it. The newly 'reconnected' field lines are highly stretched, the magnetic tension pulls fluid outwards. Magnetic stresses are converted to KE of the flows. The lower pressure in the sheet then leads to inflow from both sides of the current sheet. Thus reconnection can be a *self-sustaining dynamical process*. Some magnetic lines of force previously connected between, say, R' and R are afterwards connected from R' to Q', the magnetic topology dramatically changed.

cycling global dynamo to operate. By analogy, we can imagine the Sun as threaded throughout by tubes like bicycle inner tubes, with pressure and tension, tied into twisted and intertwined bundles. If stresses build up enough in the tubes, before something else happens, eventually something must break. Dissipation—the melting of inner tubes into one another across contact surfaces—may work to build a credible dynamo. The generation of contact surfaces can in principle allow resistive decay of magnetic fields to take care of the build up of stresses. Such a mechanism might allow the stressed flux tubes to *relax*, perhaps to evolve to simpler configurations from which Ω- and α-effects can begin,

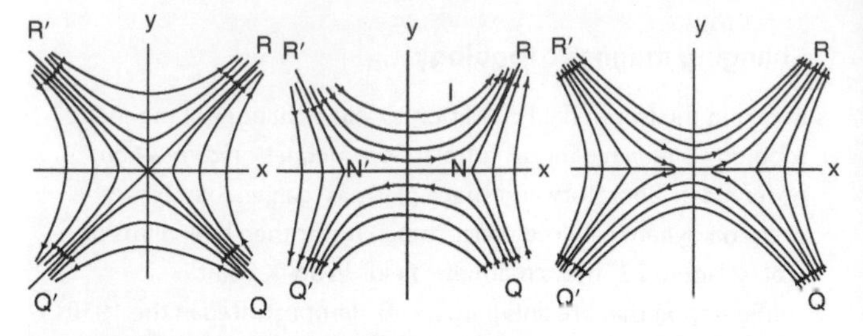

27. Left: A configuration in which the magnetic field changes sign between the upper and lower halves of the figure. Centre: An electric current flows in to the plane of the page. The current forms a 'sheet' along the x-axis by the squashing together of the upper and lower halves of the magnetic field in the left panel. Right: Resistive reconnection of magnetic field across the current sheet $N'N$, leaving the two highly stressed magnetic fields to pull the fluid outwards along the x-axis.

again and again on time scales of a decade. The Ω-effect is a large-scale phenomenon, winding magnetic fields around the relatively large solar core. But the α-effect is on the scale of cyclones, or perhaps active regions, and these scales are smaller. So we can see that the $\alpha-\Omega$ dynamo concept not only can complete a magnetic cycle, but it does so by coupling large and small scales.

This concept illustrates what might be needed for the Sun to find *some* way to couple large to small scales to permit global changes in topology. Dynamical *magnetic reconnection* makes the topology change more rapidly than simple diffusion, allowing global-scale interior magnetic fields to relax. The dynamical time scales are very fast in the solar corona, perhaps explaining the origin of solar flares. They are considerably slower in the dense interior, but nevertheless they can relax stresses built up in magnetic fields generated by the dynamo.

The above discussion contains the elements of a broad variety of dynamo models. Until 1987, models were free to assume various forms for the solar internal rotation that generates the Ω-effect. However, major breakthrough came in 1987, when the internal rotation of the Sun was measured for the first time, permitting the kind of arguments put forward by Parker and others to be challenged with real data. Along with this achievement, scientists were able to measure the internal thermal structure of the Sun. Experiments consistently showed neutrino counts 1/3 of those predicted by solar models compatible with the measured solar structure (see Chapter 1). It is now generally accepted that the neutrinos have a finite mass, permitting them to change their flavour on the way to Earth (they are generated in the 'electron neutrino' quantum state at the Sun, a state that then changes before arriving at Earth) thereby resolving the discrepancy.

Both of these breakthroughs were made possible through *helioseismology*.

Helioseismology

Before 1987, essentially nothing was known about the internal rotation of the Sun. In Chapter 3 we saw that the solar surface rotates *differentially*, with a period near 24.5 days at the equator, extending to thirty-seven days at the poles. But then, new data permitted researchers to delve into the Sun's *interior rotation* through a remarkable technique based upon *seismology*. This relatively recent birth of solar seismology is not surprising. While origins of earthquakes were pondered in antiquity (Thales of Miletus, *c.*585 BCE), it was not until the early 20th century that earthquake studies became quantitative. Studies of elastic properties of rock, new quantitative data of earthquakes, and theoretical works were needed before seismology could advance. By 1926, Harry Fielding Reid (1859–1944) had the tools and data necessary to propose that the core of the Earth is largely liquid. By 1937, a solid core within the larger liquid core was identified by Danish scientist Inge Lehmann (1888–1993). Both studies were based upon seismology, and both are important for the Earth's dynamo.

Solar (helio-) seismology has an interesting history which is still being written. The Sun is elastic—it supports many kinds of wave, including the terrestrial 'p-waves' or pressure (sound) waves, where compression and rarefaction propagate at a speed determined with the fluid's inertia. Unlike Earth's mantle, it has no shear waves ('s-waves' in the terrestrial literature) because there is no equivalent to the strong stresses in planes perpendicular to the wave direction.

Helioseismology began through measurements of oscillatory solar surface motions made during the 1960s. Measurements of narrow spectral lines trace with high precision the speeds of the solar surface. Just as for police radar, the Doppler effect detects speeds along the direction towards the observer. The Sun oscillates mostly with periods between three and ten minutes, peaking near five minutes, lasting twenty minutes at any particular location. The oscillations were found to be grouped in small patches on the surface, with a diameter a few per cent of the solar radius.

The nature of these motions remained a puzzle until 1975, when Deubner showed that most of the observed motion existed in a kind of 'harmonic' geometric pattern. Such patterns were predicted in 1970 by Roger Ulrich (1942–), who had developed a model showing that the Sun should produce a very particular harmonic structure. Just as a piano, guitar, or other instrument produces a recognizable pitch, the Sun oscillates only at specific frequencies for a given length scale. Musical instruments are designed to produce mainly just one note. Deubner's analysis conclusively showed that the solar oscillations are more ordered than, and not directly related to, the more vigorous and random convection. Unlike a musical instrument, the Sun has no special length scale in which the oscillations are 'tied at the ends'. Instead, the waves are trapped in a 'wave guide' bounded at the top by a reflective region in the visible atmosphere as gravity plays a significant role in the motion of fluid, and at the bottom by refraction which turns a downward-moving wave into an upward-moving wave. The waves then return to the surface, often many times, before they get disturbed by convection and lose knowledge that they are really part of a wave at all. The key point is that *these motions, seen only at the surface, encode information about conditions in the interior.*

The physics of piano wire- and global solar-oscillations is entirely analogous to quantum mechanics. All involve wave equations, and in each case the 'quantization' (selection of particular frequencies) occurs because we impose conditions on the sound waves (fixed points, or 'nodes') and the wave-functions (going to zero at infinity) in quantum physics. The solar waves can be broken down into individual 'modes' (sound harmonics, 'eigen-functions' in quantum theory) with certain quantum numbers. Those familiar with music will know that a piano wire fixed at each end oscillates with an integer number n of half-wavelengths. Recall that a full wave motion traces out something like the letter 'S' on its side—a half-wavelength is just one half of the 'S'. On a piano there is just one quantum number n because there is only one direction of wave propagation (along the wire). The fundamental has $n = 1$, first harmonic $n = 2$, second $n = 3$, etc. In 1979, individual modes were finally seen as the individual building blocks within Deubner's patterns, by using very long periods of observations. When oscillations are measured for longer periods, it is easier to identify their precise 'pitch' or frequency.

There are in principle three quantum numbers in a nearly-spherical elastic body like the Sun, n, ℓ, and m, because the fluid can move in three independent directions. By studying the properties of waves as functions of n, ℓ, and m, the helioseismologists have provided information critical to Parker's original picture and dynamo properties in general. Only in and after 1989 were data obtained and analysed with sufficient information content to permit meaningful studies of differential rotation. The procedure of finding and interpreting mode properties is an example of an 'inverse problem', and such problems are often 'ill-posed'. For example, one might desire to find the most information from an earthquake from a small number of observations. Common sense would suggest that if one has say three truly independent measurements, one cannot try to extract more than three independent pieces of information from the data given, without additional information. While there are natural limits to what we can learn from helioseismology, the results are clear, reproducible, and persistent. An example is shown, like contours on a map, in Figure 28, except that contour line traces a constant angular speed around the rotation axis of the Sun, instead of heights of a mountain. There exist large regions where the contours are very close. These are places where the speed changes rapidly with radius and/or latitude. These regions are precisely those places where the Ω-effect will stretch magnetic field lines out and generate toroidal field from poloidal field. Helioseismology set the first constraints on the efficiency and nature of this effect in the solar interior. These results represent probably the most recent real breakthrough of fundamental importance to solar physics.

There are three identifiable regions of strong speed gradients, or 'shear':

- a steep *radial shear* layer at the interface between the radiative and convection zone, at high and low latitudes. This thin region, located at 72 per cent of the solar radius, has been called the 'tachocline';
- a *latitudinal shear* at all latitudes throughout the convection zone. This pattern *extends to the observable solar service*, that is, helioseismology suggests that we can see the latitudinal shear throughout the convection zone by observing the surface, except for the effects of the last region:
- a *radial shear* layer near the surface of the Sun.

Much attention has been brought to the tachocline region as a place where the Ω-effect must happen. This feature lies between the radiative core and the convection zone. But it should be noted that the other sheared regions might be important for a global-scale dynamo.

Now since helioseismology is an inverse problem, it is limited in its ability to retrieve information from measurements. In particular, properties of small-scale features on scales below roughly 5 per cent of the solar radius are not retrievable. Parker's appeal to small-scale cyclonic motions remains something beyond our current ability to test through observation. Thus, even with helioseismology, we are left with only partial observational constraints on our understanding of the global solar dynamo.

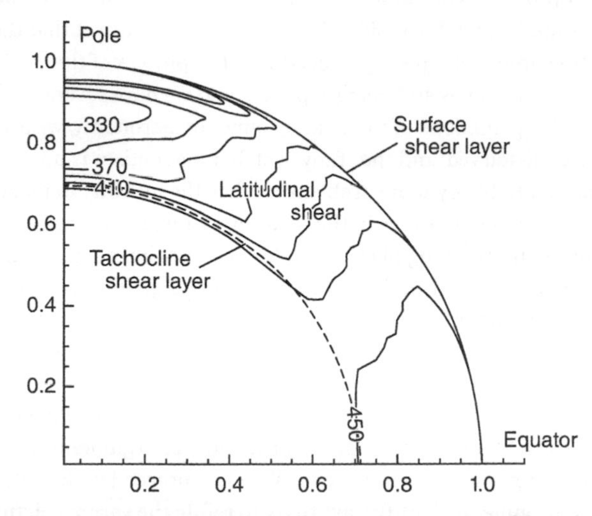

28. The internal rotation rates *in the toroidal direction* of the solar interior as a function of radius and latitude as derived from helioseismology. The lines are contours of constant rotation rate in units of 10^{-9} Hz. They are akin to the height contours of a map, the larger the rotation rate, the higher the mountain. Steep slopes in the figure are favourable locations for the amplification of magnetic fields through the Ω-effect.

Computer simulations of the global solar interior cannot probe this problem for almost the same reason: there is simply not enough room in a computer memory to solve accurately for the dynamics on all scales. The highest resolution numerical experiments hint at possibilities but currently tend to open more questions than they can answer. The same is true for global models of the Earth's atmosphere—one must choose between local models (weather) or global models in which some of the physics and chemistry of weather is replaced with a suitable prescription (these are called 'sub-grid-scale models').

Current status of solar dynamo models

Physicists have constructed a diverse set of models to explain salient properties of the solar magnetic cycle. Each kind of model has its merits and down-sides. There is no consensus yet that the solar dynamo is completely understood. The physics of the tachocline region is still being explored, and we have several physical explanations of the solar dynamo. In astronomy, we are often 'data-starved' and the study of stellar magnetism is no exception. In the dynamo problem, we face the inability to make truly critical observations in the solar interior where much of the 'action' is surely taking place. Therefore we are obliged, to make any progress, to fold-in theoretical notions and quantities with our analysis of pure observation. So it is not always straightforward to use 'observation versus prediction' to refute hypotheses, to discriminate between science and pseudo-science, as famously argued in the mid-20th century by the philosopher Karl Popper (1902–94). More modern views regard Popper's arguments as useful, but perhaps too restrictive. We must proceed with caution and some guile to identify ways to try to refute the various pictures we have built. One interesting observation comes from observations of magnetic activity and solar-like activity cycles in cooler, red stars. These are convective throughout their volumes, with no radiative zone and therefore no Sun-like tachocline. Yet red dwarf stars often harbour enormous and strong star-spots.

This all remains an exciting research area. New measurements of the Sun with observatories including the Daniel K. Inouye Solar Telescope (DKIST), the Parker Solar Probe, and of stars with ground-based global telescope networks and with the Transiting Exoplanet Survey Satellite (TESS) mission will ensure steady progress as new models are tested against ever more critical observational tests.

Chapter 4
The dynamic corona

One of the remarkable properties of our star is its tendency to emit plasma and magnetic fields into space. This 'dynamic' behaviour results from the ever-changing magnetic fields discussed in Chapter 3. Figure 29 shows a dramatic example of coronal dynamics, a large coronal mass ejection (CME) that occurred on 7 April 1997. These CMEs are common events, on average our instruments can see a CME about once a day. Near solar sunspot maxima, the Sun emits roughly three CMEs a day; at sunspot minima a CME is observed every five days. The average mass of a CME, determined from their brightness, is 10^{12} kg. The Sun's mass is 2×10^{30} kg, Mt Everest has a mass of roughly 10^{14} kg. The corona is also dynamic between these ejections: it sustains a wind; and it also readjusts its form in response to forcing of magnetic fields from beneath.

Humanity first became aware of the solar corona through total eclipses. The typical corona seen during total eclipses is a million times dimmer, but several hundred times hotter than the visible surface. It is as bright as the full Moon. Today it is easy for us to marvel at a total eclipse through the eyes of a society informed by the scientific enlightenment. But before the phenomenon was understood, it instilled awe, reverence, and even fear into the hearts of humanity, as expressed in the art of cultures worldwide.

Visual observations of total eclipses from antiquity, documented by Plutarch (*c*.46–120) and Philostratus (*c*.170–247), clearly describe the extended glow around the lunar disc which we recognize today as the solar corona. Its origin (solar, lunar, or even atmospheric) remained unknown until scientific expeditions in the 19th century revealed clues. Photography had begun in the first half of the 19th century. The first useful photograph of the corona—a daguerreotype—was made by Prussian photographer Julius Berkowski at the Royal Observatory in Königsberg, Prussia, on 28 July 1851. Observations of the eclipse of 1869 from different sites revealed that the corona belonged to the Sun. It was not a lunar, optical, or terrestrial phenomenon. It was seen to be self-luminous in a bright line in the green region of the spectrum, and the Moon was known to have no detectable atmosphere. Any remaining doubts were removed by photographic triangulation methods employed during the eclipses of 1870 and 1871, revealing that the corona showed

> the same details of coronal form and structure, and are, by themselves, sufficient to demonstrate that the main features of the phenomenon are independent of our terrestrial atmosphere and the accidents of the lunar surface.
>
> Charles A. Young in *The Sun* (1892).

Janssen in 1879 drew similar conclusions from the stability of the corona seen during the 1871 eclipse. The association of the corona with magnetic fields was suggested by the curved ray-like structures, which reminded observers of eclipses of the lines of magnetic force traced around laboratory magnets. By 1871, coronal features were already suspected to be correlated with Schwabe's decadal variations in sunspot numbers. But at the time, the nature of sunspots was also unknown.

These events occurred during an era of rapid and broad scientific advancement, yet the intrinsic nature of the corona remained a mystery. The use of the spectroscope in this period revealed unique, unidentified emission lines from the corona during eclipse. Two lines were particularly strong: the 'green line' at 530 nm and the 'red line' at 637 nm. Further advances followed World War One. In the late 1920s, Lyot (he died in 1952 in Cairo on his way back from a successful total eclipse expedition) developed the coronagraph, an instrument to see the corona by making a fake eclipse actually inside a telescope. Meanwhile, Walter Grotrian (1890–1954) obtained spectra of the corona during the 1929 eclipse. He not only confirmed earlier work which showed that the spectrum consisted mostly of scattered photospheric light, but he realized that the strongest of Fraunhofer's photospheric absorption lines could be observed in the corona, noting that they were extremely *broad*. Grotrian suggested that these Fraunhofer lines might be broadened through the Doppler effect, if the random speeds of the 'fog' of scattering particles were $\approx 7,500$ km/s. Unhindered, a particle moving this fast would cross the Sun's diameter in 90 seconds. Maxwell's theory of radiating charges showed that particles of small mass scatter radiation most efficiently. The scattering of light by the corona therefore was most likely scattering by electrons, and the smearing required an electron temperature just over 1 million K. With new coronagraphic measurements, Lyot suggested that the spectral width of the green line might be due to thermal motions, if the radiating ions had temperatures of several million K. But the elements associated with these mysterious new lines remained unknown.

On the eve of of World War Two, Grotrian proposed tentative identifications for the red coronal line (Fe^{9+}, an iron atom with nine electrons removed) and a line at 789.2 nm (Fe^{10+}). Finally, in 1943 Edlén identified four lines against laboratory measurements, and other lines using wavelengths estimated from regularities in the energy level structure of highly charged ions. Most solar spectral lines are 'allowed' by certain rules related to symmetries in quantum mechanics. But these new lines were 'forbidden' by these usual rules, allowed by other, weaker

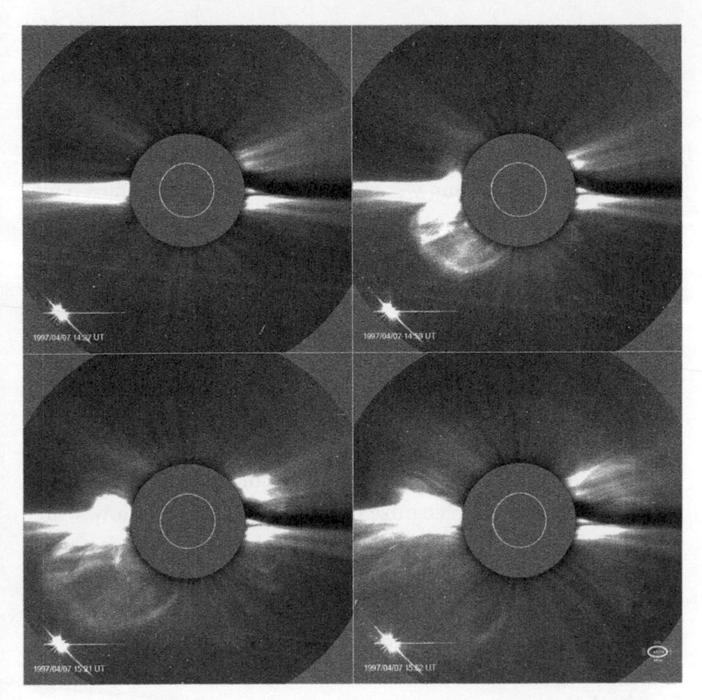

29. A time sequence of four white-light images obtained on 7 April 1997 with the LASCO instrument on the SoHO spacecraft. The middle two frames (top right, bottom left) show clearly the CME emerging from the lower left quadrant of the Sun.

physical interactions. These were transitions of ions with between nine and fifteen electrons removed. The green line was identified as a forbidden transition of Fe^{13+}. The energy needed to remove thirteen electrons from iron could be estimated using quantum mechanics as well as laboratory data: it is about 1,000 electron volts. In the photosphere, the average energy of photons and particles is 0.4 electron volts. Using Boltzmann's statistics, we find that about 0.001 (0.3) per cent of electrons heated to one (two) million degrees have energies above 1,000 volts. Yet this fraction is sufficient to account for the presence of such ions in the tenuous corona.

So, by the end of World War Two the stage was set for the beginning of coronal physics. The only conclusion had to be that the outer parts of the Sun were much hotter than its surface. This made physicists very uncomfortable, and it took a little time to be accepted. The reason? Everyone 'knew' that the centre of the Sun must be hotter than the outer layers. Our experience tells us that heat moves from hot to cold and not vice versa. It was as if, sitting in front of a coal fire, the temperature of the air exceeded that of the fire. This non-sensical experience is expressed in physics as the Second Law of Thermodynamics.

The precise origin of the corona remains unknown to this day. The key to the puzzle is to recognize that while *heat* cannot flow, with any meaningful probability, from hot to cold, *energy* can, in the right (ordered) form. So, if just a small amount (one part in 100,000) of the Sun's thermonuclear power supply can be converted to some ordered mechanical form and dissipated appropriately above the surface, then the problem would be solved.

Why bother with the corona?

There is on average little mass or energy in the corona, steady solar wind, and CMEs. So why is the corona of interest? First, it presents us with a natural laboratory for studying plasma physics of unique relevance to laboratory physics, nuclear fusion devices, and the physics of other astronomical objects. Next, while the *quantity* of energy needed to sustain the corona is almost negligible, the *quality* of this energy must be high, also varying with magnetic activity. Thermodynamics tells us that heat is the most disordered, lowest quality form of energy. In order to maintain the corona, energy is needed in an ordered form, such as directed mechanical or electromagnetic energy. The precise nature of the processes responsible for coronal heating has eluded our best detectives for eight decades, in contrast to the many advances made in other fields, such as genetics. As a result of this heating, the corona converts a small but important amount of the Sun's core nuclear power into variable UV, EUV (extreme ultraviolet), X-rays, radio, and, during flares, γ-rays, and accelerated particles. Without the storage and release of the free magnetic energy within the corona, the Sun would produce none of these high energy outputs, and the Earth would not possess its upper atmosphere, particularly the thermosphere and ionosphere.

The corona continuously emits hot plasma (the solar wind) and sporadic magnetic field bubbles ('plasmoids') into the near-vacuum of interplanetary space. The wind causes magnetic braking, slowing the Sun's rotation and weakening the solar dynamo during its time on the main sequence. The corona can shed *magnetic helicity*, a topological quantity generated by the dynamo that would otherwise accumulate and stifle cyclic behaviour. In this sense, the corona acts as a valve, releasing the 'waste product' (helicity) into space. Without the coronal valve, the cycles of sunspots would cease. With it, the Sun is obliged to shed plasmoids, including the enormous CMEs, into space.

Salient coronal properties

The features in the left panel of Figure 30 are magnetic fields rooted under the photosphere, measured using the Zeeman Effect applied to photospheric spectral lines. On top of these concentrations are arch-shaped coronal 'loops' originating from them and inter-connecting them, and coronal 'fans' that either return to the Sun or extend into space. The figure can be compared with the sketch of Figure 11 in which smoothly structured coronal magnetic fields overlie tangled magnetic fields below, as the force balance changes with height above the surface, from hydrodynamic to magnetic dominance. The arches and fans

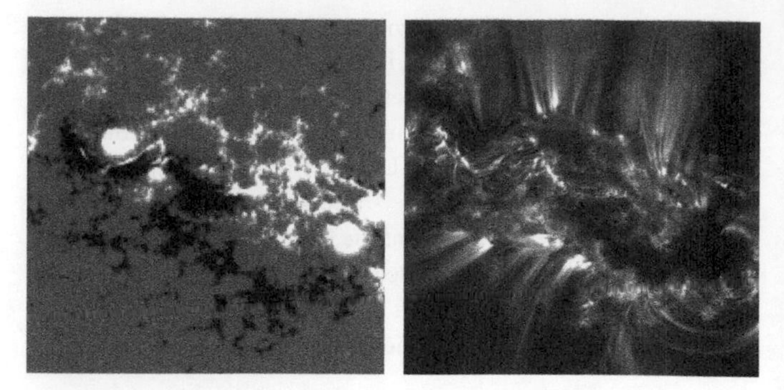

30. The left panel shows magnetic fields penetrating the solar surface from a large active region, the right panel shows an image of the overlying coronal plasma in the light of the ion Fe^{8+}. The left panel is from the Michelson Doppler Imager instrument on the SoHO spacecraft, the right from the TRACE satellite. The images were obtained on 24 April 2001 within 20 minutes of one another. The entire region is about 300,000 km across, an area 1600 times that of the Earth.

are smaller-scale versions and subsets of the structures seen in white light during eclipse (Figure 10). Those in Figure 30 show only plasma at temperatures near 1 million K, unlike the white light eclipse data which show all the coronal plasma. In the late 1970s, work with SKYLAB data revealed that each of these coronal features can be usefully thought of as a narrow strand of plasma, each of which, owing to Alfvén's theorem, belongs to an individual rope of magnetic field. In this picture, some important physics of the three-dimensional structure of the corona can be reduced to the one dimension running along the rope. The physics behind the idea is based on Alfvén's theorem and on the fact that heat conduction dominates energy transport in the corona, along magnetic field lines. In the mid-1950s, Sydney Chapman had shown theoretically that coronal material near temperatures of 1 million K must be a very efficient conductor of heat. This picture is still useful for coronal research studies today.

Outside of eclipse, the chromosphere, transition region, and corona are visible only at special wavelengths. EUV and X-ray wavelength bands have been used for coronal work on many missions. The Sun's disc is dark at these wavelengths because the radiation emerging from photosphere and chromosphere beneath comes from cool plasma below 10,000 K. UV, EUV, and X-ray images have been regularly obtained by instruments on the SKYLAB, SMM, SoHO, TRACE, STEREO, SDO missions, and many sub-orbital rocket flights. Figure 30 shows magnetic data from the Michelson Doppler Imager on SoHO, and EUV data from TRACE. Each fan and arch tends to be at the same temperature because of the dominant effects of heat conduction at coronal temperatures, pointed out by Chapman.

In the six decades since we recognized the corona as a million degree plasma, we have identified some basic properties which must be accounted for by any successful theory. Coronal heating must occur through dissipation of ordered free energy associated with magnetic fields. Outside of active regions, only about one part in 100,000 of the Sun's energy is needed to sustain it. This figure is roughly the same for the quiet Sun and coronal holes, because although the holes are dark, much of the energy from heating is used to accelerate the plasma into space. The dissipation must occur on very small scales, because the fluid is highly conducting and inviscid. The persistence of the coronal plasma indicates that the heating mechanisms must be steady in time, or they must occur in bursts separated by at most half an hour; if not then the plasma would be too cool.

Over sunspots and active regions, we can add the following: coronal heating is more intense where the underlying magnetic field is stronger, yet it is intense only in some of the magnetic ropes extending above the surface. Coronal temperatures are higher over the core of the sunspots, and lower on the periphery.

An unsolved problem

In trying to solve the coronal heating problem, ideally we would observe the injection of magnetic energy from the surface into the corona, follow the energy upwards and observe it in the act of dissipation. But this direct approach faces many practical problems. We can measure magnetic fields in the photosphere fairly well. Sometimes we can do it also in the chromosphere, if we are lucky. But we have yet to measure magnetic fields in the corona with any kind of precision or with the needed detail.
The observable signatures are simply too weak. The challenges involved in sleuthing this problem are deep, diverse, and devilishly difficult, perhaps worthy of the attention of Conan Doyle. Instead of catching the culprit red-handed, we are left, like Holmes and Watson, looking for circumstantial evidence. Inspector Lestrade of Scotland Yard would declare, 'But Holmes, we know exactly who the culprit is!' It is the changing magnetic field. What we don't know is the method used, and the culprit has so many options and dirty tricks to fool us. We don't know if we are looking for a smoking gun, a lead pipe, a rope, poison, or something else. Worse still, physical processes clean up and confuse the scene of the crime, hiding the evidence from our telescopes. Why is this the case?

First, just one part in 100,000 of the Sun's power is required to sustain the corona. A minuscule change in conditions in the photosphere can lead to enormous changes in the corona. We can say that the problem of coronal heating is *ill-posed*, in that unobservably small changes in photospheric magnetic fields can lead to enormous effects in the corona. To illustrate the problem, a typical car engine generates 100 kW of power. The number plate lights require about 10 W, one part in 10,000 of the engine's output. Searching for causes of coronal heating by looking at changing photospheric properties is like trying to see if the number plate lights burned out by measuring the associated increased speed of the car.

Second, even if we could observe the small changes in upward flux of energy into the corona, the *dissipation mechanisms* which convert them into heat must occur on *tiny, unobservably small scales.* This is because of the high electrical conductivity (and small viscosity) of coronal plasma. Just as a copper wire will heat and melt only under very high electric currents, this energy dissipation requires enormous currents densities, j, which necessarily imply changes in the structure of magnetic fields on tiny physical scales which are currently beyond our ability to measure.

Third, once heated, heat is conducted along magnetic field lines so rapidly by electrons that it is ducted far from the crime scene—the energy dissipation site. By the time we observe any radiation, it has been distributed elsewhere to be converted to KE of motions and/or radiation in plasma at lower temperatures. It is as if the culprit, after committing the crime, has re-arranged the scene and removed nearly all the evidence of fingerprints, footprints, and so forth. Additionally, the chromosphere/corona interface is thermally complicated (see Figure 8). The transported coronal heat can be mistaken for heating directly in the chromosphere, or simply masked by other structures along the line-of-sight.

Fourth, unlike daily experience where more heat generally leads to a proportionately higher temperature, electron heat conduction moves heat to cooler plasma, which radiates energy with far greater efficiency than the corona. Temperature changes *within the corona* can therefore be below our ability to measure, even for large changes in heating.

Last, not all of the magnetic energy that makes it into the corona is 'free' to do work or heat anything. In fact, under typical coronal conditions, magnetic energy dominates over the energy in the plasma, but most of the energy is not of the 'free' variety. Most of the emerged magnetic field is generated by strong sub-surface electrical currents, themselves determined by the sluggish effects of slow stretching and twisting motions under the surface. So, even if we could measure coronal magnetic fields, we would have to measure even smaller changes in the fields to look directly at how magnetic energy is converted to heat. When we look at images such as in Figure 30, we must conclude that we are almost blind to the magnetic free energy that heats the corona. Until we can measure coronal magnetic fields in detail, we cannot observe directly the conversion of free energy to heat. It would be like trying to explain what we see at a sheepdog trial without the ability to see the dogs! Worse still, only some free energy is available for heating. Rogue dogs are allowed to disrupt the trial, sending the sheep away from their intended gates.

A way forward

The difficulties facing researchers attacking the heating problem are in essence that we are data-starved, and that the data we have are not directly linked to the processes we are trying to understand. In this situation, the only choice is to add information. The new facilities (DKIST, Parker Solar Probe, Solar Orbiter, discussed in Chapter 6) will open genuinely new windows into the problem. There is some hope that the combination of *in situ* measurements within the corona with measurements from DKIST can shed light on the coupling across scales, for example.

But given the nature of the challenges, it seems wise to add yet more information, where possible. We have a firm physical basis upon which we can build conceptual, theoretical, and numerical models of heating processes. One approach is to *develop models of various processes and test their predictions against data*. This is the method most frequently used today. If a given model survives critical tests from data and is based upon sound physical principles, it remains a candidate.

One class of model was successfully eliminated. Several studies have quantified the available energy carried mechanically, in

Waves

Wave motions happen when changes in inertia are balanced periodically by a 'restoring force'. A pendulum weight can be lifted and let go. The work we do against gravity is stored (potential energy) and upon release the weight is forced by gravity in an arc, gaining KE at the bottom, and then again moving upwards to store potential energy. Sound occurs when air is pushed forwards compressing the air in front. The KE is stored as internal energy as a local increase in pressure, the increased pressure pushes neighbouring fluid, converting internal energy back into KE. In MHD there are two additional stresses that can store potential energy: magnetic pressure and tension. In MHD we have three wave 'modes' characterized by how KE (inertia) is shared alternately with potential energy (restoring force), called slow, fast, and intermediate modes depending on how fast the oscillations propagate. When magnetic stresses dominate, the slow mode is like a sound wave moving along tubes of magnetic flux. The fast mode moves almost the same in all directions. The intermediate ('Alfvén') mode is quite different and does not compress the fluid. It is a wave in which motions are perpendicular to the magnetic field lines: this can be a simple waggle like a dog's tail, or a rotation around the magnetic lines of force. The intermediate mode propagates along the direction of the magnetic field with a speed called the 'Alfvén speed'. Just as the sound speed varies as $\sqrt{p_{gas}/\rho}$, where p is the pressure and ρ the mass density, the Alfvén speed varies as $\sqrt{p_{mag}/\rho}$ or $B/\sqrt{\rho}$, even though sound is compressive in nature and the Alfvén mode is not.

sound waves, and compared this with the energy required. In 1978, R.G. Athay (1923–2015) and O.R. White (1932–) demonstrated, with the help of a model chromosphere, that sound waves fell short of heating the corona. But can't sound waves be eliminated by noting that the corona is structured by the magnetic field? After all, pure sound waves arise only from changes in gas pressure and inertia and know nothing of magnetic fields. Not at all. In the presence of magnetic fields, a fluid supports a slow wave (modified sound wave), an intermediate wave, and a fast wave. Athay and White eliminated the slow wave, which propagates along tubes of magnetic flux in the corona. The work is important because it demonstrates *refutation* of an entire class of physical model. It might even have pleased Karl Popper.

The literature on magnetic heating is vast, but there are common threads. A central theme is to find ways to drive the system to the smallest scales needed to dissipate energy. Mechanisms for transport and dissipation are usefully put into two categories: waves propagating along the emerging magnetic flux ropes, and interactions within and between magnetic flux ropes. This distinction is physical. The free energy in waves is ephemeral, shared between the wave motion and magnetic pressure and tension. In the interaction picture, the energy is stored slowly in the overlying corona by a build up of stresses from beneath, later to be released *in situ* or even transported downwards.

Within the wave picture is an abundance of idealized MHD models. A central idea is to generate steep gradients in the magnetic and velocity fields. Shock waves such as those at the bow of a supersonic aircraft do this, but the kind of shocks that might heat a rope in an active region, flowing along the rope, are of the slow kind. These are in essence the same as the sound waves that were discounted in 1978 by Athay and White. The fast modes do not follow the magnetic field and so to account for the kind of observed loop or rope structure seen in Figure 30, fast-mode energy would have to be preferentially dissipated into these loops. Mechanisms to do this exist in principle, but this idea has not received much attention, perhaps because it presumes the existence of a loop pre-conditioned to accept this energy, and the mode by itself cannot do this. There remains then the intermediate mode which is akin to a wave travelling along stressed wires. Let us follow this analogy and imagine that an observed coronal loop is essentially a conglomeration of different ropes melded together, each sub-rope having its own density and Alfvén speed. This is expected—the plasma in the ropes must come from below, where the chromosphere is thermally inhomogeneous (e.g. Figure 9). Now send intermediate waves along the sub-ropes from beneath, by moving the ropes from side to side or rotating them, at a given frequency. Even under these ideal conditions, very sharp gradients are expected to arise in the magnetic fields *between* the sub-ropes as the waves march forward at different speeds. Even if the wave motions remain linear, the differences between them can grow without limit, until a physical process such as viscosity or electrical resistance dissipates the associated energy.

This process is called 'phase mixing'. An analogy would be to take two or three piano wires of different thicknesses and lengths, put them under the same tension, and pluck them at the same time. Left to their own devices, a chord would be struck. But now bring them together so that they are physically touching. One can imagine a cacophony of sound as the friction between the wires starts to dissipate the wave energy. The wires heat up, the chord previously heard quickly dies away. In MHD, the friction can be of almost the same form, the collisions between electrons and protons lead to electrical resistance, electron–electron collisions lead to heat conduction, and proton–proton collisions dominate momentum transfer leading to fluid viscosity. Collisions that take ordered energy and turn it into random particle motions are the quintessence of the dissipation of ordered energy forms into disordered forms, that is, heating.

Phase mixing lies at the heart of wave-heating models. It works so long as the wires are free to oscillate separately. It works along both the magnetic fields open to space as well as those returning to the Sun. For the latter, there is an amplification mechanism that occurs when the waves are tied at both ends like the piano, and the wave 'resonates' as a natural mode, a 'standing wave', when waves reflected at each surface area reinforce one another. A layer of rapid wave energy dissipation is set up when the fluid motions coincide in wavelength and frequency with the Alfvén speed, which is expected to vary within a loop containing plasma at a variety of densities. This process is called 'resonant absorption'.

In the real Sun, in which magnetic fields at the surface and above are anything but tidy, such effects seem inevitable, and because the Alfvén speed exceeds the sound speed, when, as in the corona, densities are very low, there is ample energy carried by such waves, so that it seems that the problem may have found resolution. But nature is not quite so kind. Parker has shown that the piano-wire picture is too simple. In the phase-mixing model, the waves are essentially *two-dimensional*, they propagate along the stressed fields, and interact only across them in one other dimension, not allowing any oscillation in a third dimension. Parker showed that by ignoring this third dimension, all of the wave energy *must* go into generating steep gradients as predicted by the model. By allowing motions in the third dimension, and including lowest order non-linear effects, Parker argued that instead of dissipating, wave energy is ducted in other modes which propagate along the interfaces.

So the debates and modelling efforts continue.

Next consider the 'interaction' picture. Since 1972, Parker has drawn attention to a remarkable and general property of highly conducting plasma in which the magnetic stresses dominate. The ratio of fluid to magnetic pressure in MHD is $\beta = p/(B^2/\mu_0)$. Here, p is the gas pressure, B the strength of the magnetic field, and μ_0 is a constant of nature called the permeability of free space. Now, over an active region, UV spectroscopy shows that $p \approx 0.1$ Pascal (Pa). But $B^2/2\mu_0$ exceeds about 40 Pa, so $\beta < 0.003$. Parker is a master of taking problems that are not soluble and, instead, solving a related and soluble problem, and then making inferences. Such things are part of the art of science. Like the teacher in the film *Willy Wonka and the Chocolate Factory* faced with Charlie's difficult arithmetic problem, he solves instead one that he can do. Parker asks us to consider the corona as a system where $\beta = 0$ and the effects of gravity are negligible, and examines the consequences. This is an odd thing to ask. This system is self-magnetizing and yet the pressure is zero. For a magnetic system to exist currents must flow. For currents to flow there must be electrical charges, but these charges must have zero pressure, that is, their temperature must be zero. You might think 'Aha! it's a superconductor', but alas, the superconducting state ejects magnetic field. Instead Parker performs a thought experiment in which we take the *limit* as $\beta \to 0$ and carefully consider the consequences. Parker has devoted an entire book to this problem. He focuses on two essential properties of this system: (1) the magnetic force is the only force on the fluid; (2) the fluid is ideal so that Alfvén's theorem is strictly in force: the topology is fixed for all time. He considered the configuration shown in Figure 31. Each of the plates represents an infinitely conducting solar photosphere connected by the magnetic field lines in the corona between them. It is an idealization of a sunspot, in which the two polarities are the plates, and the magnetic field between them is a coronal loop (Figure 30) that has been straightened out. The left panel shows the situation at time $t = 0$, when the magnetic field is just like a field far from a bar magnet. We could add any plasma we liked between these plates and, if it were, say, in hydrostatic equilibrium in which the weight of the plasma balanced an upward pressure, nothing would happen.

Now suppose that at the ends of the plates we allow the fluid to move around, just as granules move around. The right-hand panel of Figure 31 shows the situation after the dense and moving granules have re-arranged conditions at the plates. Parker asks: what kind of physical state can we expect for the fluid sandwiched between the plates? The answer is quite unexpected. Unless the motions in the plates are unreasonably well-ordered, there is no *continuous* solution available to the system that can satisfy the conditions of a topology fixed through the infinite conductivity and force balance. The two constraints—force balance and fixed topology—*over-determine* the system; they cannot, in general, be satisfied at the same time. The conclusion seems inevitable, this system must evolve towards states containing *discontinuities*. The only solutions possible are those having *tangential* discontinuities, that is, embedded in the solution sketched in the right-hand panel are sudden changes of *direction* of magnetic fields adjacent to one another.

On the Sun, a genuine discontinuity is neither realistic nor necessary for rapid heating to occur. Instead, non-ideal effects such as dissipation through electrical resistance or viscosity and/or instabilities will kick in. In any event, *ideal MHD causes its own demise*. Parker argues that the free energy stored as the corona tries to find a solution compatible with the frozen field and force balance will be released by non-ideal processes of heating and reconnection. The bursts of heating that can arise when a certain limiting gradient arises across magnetic ropes of flux Parker calls 'nanoflares'. This picture of nanoflare heating continues to vie as a credible mechanism for heating the corona. William Shakespeare might have been pleased with the irony that the very purity embodied in ideal MHD, when faced with nature, naturally leads to its own demise.

This conclusion has proven difficult for some physicists to swallow. After all, solar plasmas are not infinitely conducting, the photosphere–corona interface involves the complications of the chromosphere, which is not force-free. Much has been discussed regarding this theorem far beyond the qualitative picture given here. But so far closer inspection of mathematics by Boon-Chye Low and colleagues, supported by highly non-dissipative

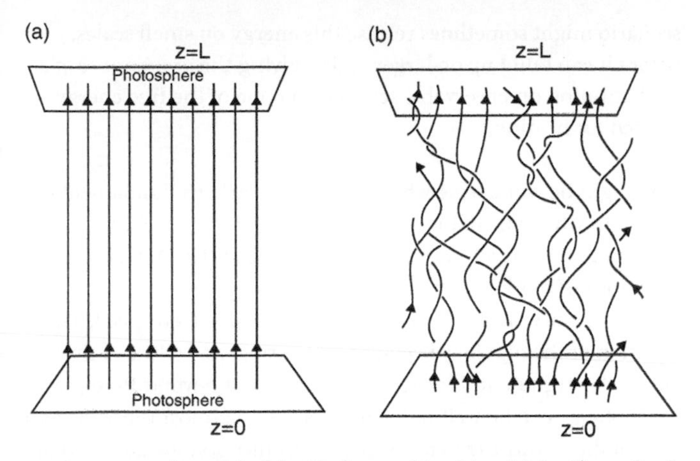

31. A conceptual model of the 'Parker problem', in which a 'force-free', infinitely conducting magnetic field rope connects two parts of the photosphere that have been straightened out in (a). Several minutes later, in (b), the random meandering of the fields caused by photospheric motions such as granules have intertwined the field, storing magnetic free energy between the two plates in current sheets. Parker showed that this system must form tangential discontinuities, owing to the inability to find a smooth solution in all but the simplest footpoint motions.

numerical simulations by Ramit Bhattacharrya (1973–), has failed to refute the theorem.

The constant but somewhat random release of free energy stored on small scales is not incompatible with the salient observational properties. It may also explain why EUV and X-ray images of the active corona so often appear to trace out magnetic lines of force which have no apparent free energy *on the much larger, observed scales*. Figure 31(b) should be compared with the twisted structure shown in Figure 35, a configuration with free energy on large scales. The large-scale free energy present tends to destabilize the structure with consequences discussed below. The nanoflare scenario might sometimes release this energy on small scales, before it can build up on larger scales, giving the *appearance* of a configuration on observable scales with none of the free energy needed for heating.

You might be wondering what happens to Parker's Fundamental Theorem when one of the 'plates' is absent, that is, the fluid is allowed to relax freely as it expands into space. The MHD equations of motion demand that the twisting motions will propagate outwards into the solar wind at the Alfvén speed. The question of the topology becomes interesting, and a 'travel time' enters the argument. If one of the plates is out beyond Pluto, say, the topology can be well-defined. This case is almost the same as a rope of flux returning to the Sun at a distance somewhat less than half of Pluto's orbit. The second scenario is identical to Parker's plate problem, except that now we must consider how long it takes for the information to propagate from plate to plate. So long as the time of the stirring and twisting of the meandering granules is shorter than L/c_A, the same arguments apply, and the system will evolve towards formation of tangential discontinuities. Using the coronal pressure scale height $L = 50{,}000$ km and $c_A \approx 1{,}000$ km/s, then all motions with periods below about fifty seconds will tend to produce tangential discontinuities (TDs) low within the corona. While below granule lifetimes, granular motions certainly exist on these time scales. It seems this outcome cannot entirely be discounted.

Some have argued that Parker's magneto-static theorem is essentially the zero-frequency limit (i.e. time independent) part of a system subject to a variety of time-variable conditions imposed at the boundaries (the plates). The generation of magnetic free energy on small scales from large scales is reminiscent of a turbulent cascade. However, the dynamics of such an MHD cascade is very different from the classical model of fluid turbulence by A. Kolmogorov (1903–87), as it applies in the Earth's atmosphere, for example.

Parker's work suggests the omnipresence of small-scale reconnection in large, untidy physical systems encountered in nature. This has important consequences throughout astronomy.

The solar wind

Evidence for particulate emissions from the Sun was suspected in the 19th century. Variations in the electrical environment of Earth were seen after the 'Carrington Event' flare of 1859, and some noted a twenty-seven-day period in fluctuations of geo-magnetic phenomena, strongly suggesting a solar origin of charged particles modulated by the Sun's rotation.

Indirect evidence for a steady wind came from the observations of comet tails—not one but at least two kinds of tails had been well-documented by observers. The cause of two types of tail ('dust' and 'ion') suggests that two kinds of momentum associated with particles and radiation emerge from the Sun.

Light carries momentum

According to Newton, the momentum of a particle is defined as the product of velocity and mass. But radiation behaves as a wave and a particle, the latter having no mass. Radiation does however carry momentum which is easily seen in action, in a 'Crookes radiometer' (after Sir William Crookes, 1832–1919), nowadays a novelty item. The device, invented in 1873, is a sealed vacuum glass bulb, with veins coated in black and white on either side attached to a low-friction spindle. The black side absorbs more radiation than the white side, heats up, and the black side emits more radiation than the white side. The emitted radiation imparts extra momentum to the black side, spinning the spindle and veins (Figure 32). The same principle is being used to develop 'solar sails' to collect solar radiation momentum in order to drive spacecraft in the solar system.

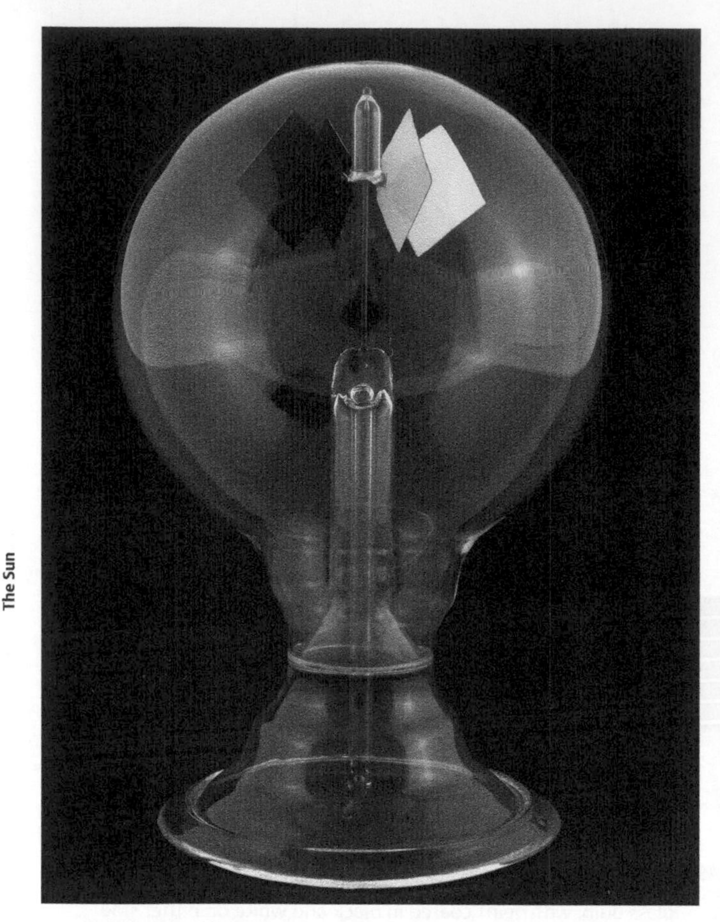

32. A Crookes radiometer.

In 1951, Ludwig Biermann (1907–86) computed how much force the radiation would have on comets. He demonstrated that radiation was insufficient, and he suggested that a flow of ions and electrons would also be needed. Plasma speeds of several hundred km/sec and densities of 100–1,000 particles per cubic centimetre could account for properties of comet ion tails. But it was not until the space age that direct evidence for a steady solar wind was obtained.

Like all fields, solar physics has its folk-lore, and a famous tale involves Eugene Parker and 1983 Nobel Laureate Subrahmanyam Chandrasekhar (1910–95). Owing to strong heat conduction, Parker argued, the corona would be hot over a considerable fraction of the inner solar system. He submitted a paper to the *Astrophysical Journal*, which Chandrasekhar served as editor. Parker argued that a one million Kelvin corona cannot be contained by the ambient pressure in the interstellar medium. Therefore, a solar wind was inevitable, if some of the magnetic fields extended from the Sun to interstellar space. Two eminent referees rejected the paper. But, as editor, Chandrasekhar stepped in and accepted the paper for publication in 1958, having found nothing wrong with the mathematics. In January of 1959, direct measurements of particles coming from the Sun were made with the Soviet spacecraft Luna 1, as it emerged from the Earth's protective magnetic field into interplanetary space. Luna 2 quickly confirmed these results. In 1962 the American spacecraft Mariner 2 sampled the solar wind on its voyage to Venus.

Parker's main idea can be appreciated by analogy with a pressure cooker. A pot of soup brought to boil emits steam once it reaches the boiling point of water, 100 Celsius at sea level. The conversion from water to steam is called a 'phase change' because it moves liquid into the gas phase. Almost all of the heat energy goes into breaking the intermolecular bonds that hold different water molecules together. The released steam exerts a pressure, it pushes the cold air out of the way. The pressure of the air is insufficient to hold the steam inside the pot. To see that the hot steam really exerts a pressure, we can add the pressure cooker's top to the pan. The steam does not, at first escape. Soon, though, the molecules inside get hotter, that is, they move faster, carrying more momentum, as this is the only place the stove-top energy can go.

The lid feels an upward force and needs to be tied down, as the steam molecules bounce off of the lid back into the pot, depositing momentum onto the lid. Eventually, the little valve that is pushed down in the lid by a spring is unable to keep the steam inside, and steam escapes, at a much higher temperature than before, out of the valve. Parker's solar wind solution has, at its core, this same effect. The interstellar medium—the space in the Galaxy between the stars—consists of gas, dust, and plasma. Although relatively cold, it does exert a pressure. But just like the pressure cooker's valve, the amount of pressure is insufficient to hold in the escaping gas. The equations that Parker solved have different solutions depending on the assumed thermal physics of the plasma, but also on 'boundary conditions', the things we have to assume outside of the region of applicability of our calculation itself. Parker showed that with the interstellar conditions at the outside boundary, the solar wind will flow, and within a relatively small distance from the Sun, the wind will be *supersonic*. This is quite unlike the pressure cooker case. It arises because of the need for the solar wind to overcome gravity.

While the pressure gradient force of a million degree corona is a major source of momentum for the solar wind, further research has shown that it is not the whole story. There are two kinds of wind, 'slow' and 'fast', that arise from physically separate regions of the solar surface. *In situ* measurements have revealed that the slow wind generally occurs closer to the equator, the fast wind near the poles. Remarkable data were acquired by the Ulysses mission, as shown in Figure 33. The spacecraft used a Jupiter fly-by to insert it into an orbit over the solar poles. The figure shows wind speeds measured *in situ* as a function of the spacecraft's solar latitude, along with sunspot numbers. The fast wind is seen clearly during the relatively 'simple' coronal configuration near solar sunspot minimum in the left panel. The two winds are spatially intermingled close to sunspot maximum (right panel), as the spacecraft's orbit intersects the two kinds of wind, revealing a mixture of large-scale streams distributed across many latitudes and longitudes.

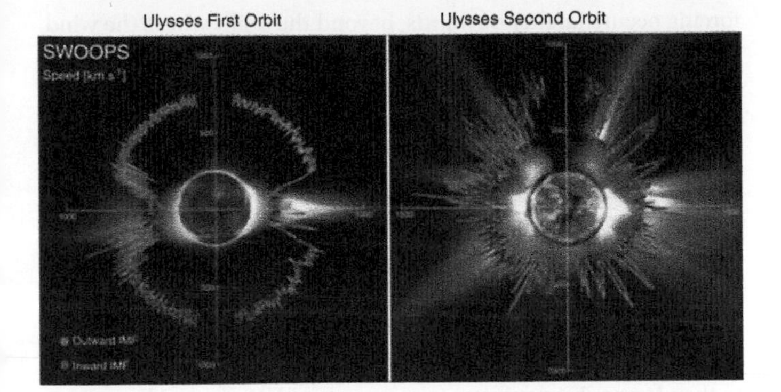

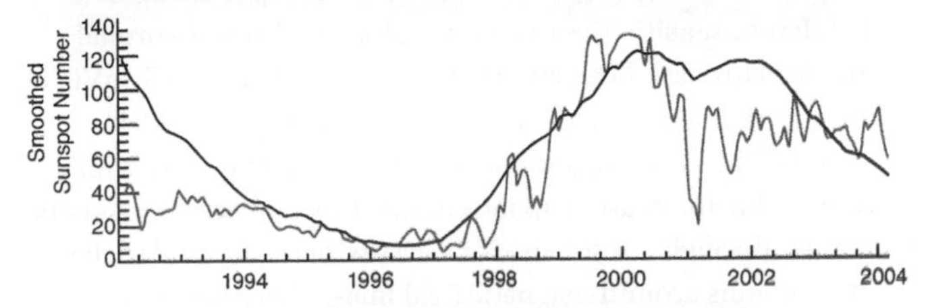

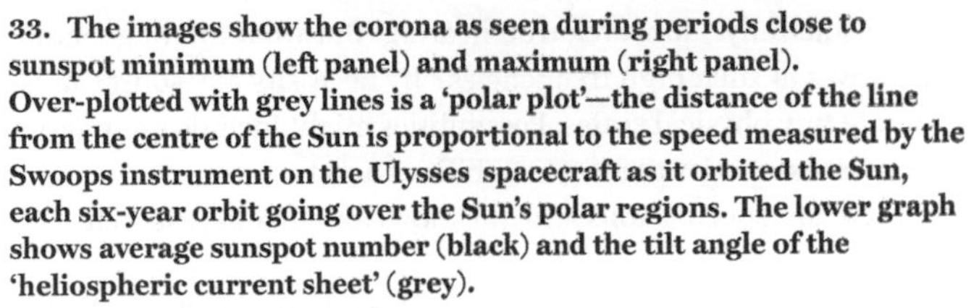

33. The images show the corona as seen during periods close to sunspot minimum (left panel) and maximum (right panel). Over-plotted with grey lines is a 'polar plot'—the distance of the line from the centre of the Sun is proportional to the speed measured by the Swoops instrument on the Ulysses spacecraft as it orbited the Sun, each six-year orbit going over the Sun's polar regions. The lower graph shows average sunspot number (black) and the tilt angle of the 'heliospheric current sheet' (grey).

X-ray imaging and extreme UV spectroscopy has revealed that the poles are darker and cooler than the equatorial regions, on average, forming coronal holes. We now associate the fast solar wind with Bartels' streams of particles called M-regions identified, using SKYLAB observations, as coronal holes. The conclusion is that the 'fast' solar wind originates from coronal holes. Models reveal that the nature of these outflows changes depending on where we choose to push the plasma outwards and heat it. Roughly speaking, when heated and forced outwards close to the Sun, the wind carries more mass outwards. If the heating and forcing occurs farther outwards, beyond the point where the wind flow exceeds the speed of sound (roughly 100 km/s), then only the flow speed is increased.

From these considerations, it is clear that the two kinds of wind require additional physics than just heating to account for their behaviour. Beyond this we know little at present, but it is an area of intense research. The 'intermediate wave' motions in the corona are capable of reaching the supersonic region near ten solar radii. For the same reason they are difficult to dissipate and thereby dump their energy into the plasma, even if momentum can be transported to the plasma. Intriguing measurements of UV spectral lines with different sensitivities to motions along and across coronal magnetic fields have been reported using data from the UltraViolet Coronagraph Spectrometer (UVCS) instrument on the SoHO spacecraft. The data suggest that kilohertz frequency waves may be responsible for heating up ions in the direction across magnetic field lines, through a resonance of waves with the natural (helical) gyration of ions around magnetic field lines. This process is called (fancifully!) ion-cyclotron heating. But what might generate such waves at such high frequencies? By necessity any source must generate tiny physical scales. Possibilities might include nanoflares, shock waves, and/or plasma instabilities and turbulence. But these are speculations and questions for current research.

The Parker Solar Probe and Solar Orbiter missions will sample the solar wind well within the orbit of Mercury, promising that this will soon be a research area of intense interest. The probe will fly within nine solar radii by 2025.

The solar wind in interplanetary space

How do the corona, solar wind, and magnetism behave farther from the Sun, in the space occupied by the planets? The corona is threaded with magnetic field lines of two basic types: those that return to the Sun and those that extend far away into the interstellar medium. The solar wind lives, by definition, within ropes constituting the latter, 'open' field lines. (The word 'open' merely means they do not close locally, in the vicinity of the solar surface, but extend into interstellar space.) A locally open condition is readily created by the change of topology, most likely dynamically through magnetic reconnection. This reconnection might happen because the thermal pressure might force open field lines, or they might open due to other forces or even spontaneously as an equilibrium is lost. In fact, there is evidence that this is happening all the time, through the rapid evolution seen at the boundaries of coronal holes.

Just as we idealized the conversion from 'poloidal' to 'toroidal' field in our discussion of the dynamo 'Ω'-effect, we can imagine building the Sun's corona starting from a purely dipolar (closed) magnetic field. A corona is made through the actions of convection, rotation, and evolving magnetism, and reconnections will change the dipolar field's topology into a different kind of field that is eventually open to interplanetary space, except where the magnetism is strong enough to hold down the plasma in loops that return to the surface. Some of the 'closed' regions straddling the equator will be drawn out by the wind into space. The dipole close to the Sun is drawn out like a candle-flame (evident in Figures 10 and 33). On the north side of the equator, the magnetic fields will be in the opposite direction to the adjacent field just south of the equator. This configuration is none other than a current sheet, a solar-sized tangential discontinuity of magnetic fields (such as shown in the middle of Figure 27). It is a narrow region of electrical current wrapped like the rings of Saturn around the equator of the Sun, except that it can become warped. It is called the 'heliospheric current sheet', and a rendering of its appearance as it extends far into the solar system is shown in Figure 34. Evidently some work done by the wind on the magnetic field is converted into electrical energy. As the Sun's magnetic cycle evolves, there may well be a superposition of several of these candle-flame structures corresponding to surface active regions

34. The heliospheric current sheet, separating southward- and northward- pointing solar magnetic fields, is shown embedded in the solar system.

and/or sunspot groups that appear on the Sun, and the heliospheric current sheet can become far more complicated.

Figure 10 shows clearly three prominent candle-flames at about 1, 3, and 8 o'clock. Figure 33 reveals two (left panel) and a confusing superposition of many (right panel). The inference is that these structures host current sheets which continue to extend into interplanetary space. When such a sheet reaches the Earth, the Sun's magnetic field can easily change sign as the sheet moves north or south of the Earth in its orbit. Such changes have important consequences in the space around the Earth, that is, for space weather.

It might be assumed that, like the spokes of a wheel, the solar wind flows, and drags along magnetic field, only in the radial direction from the Sun. But the Sun is rotating, and although the magnetic field does have some stiffness (the magnetic stresses) that will enforce a radial expansion close to the Sun (see the rays of coronal light from the north and south poles of the Sun in Figure 10), farther out the plasma inertia overwhelms the stresses and the magnetic fields form a spiral structure. Such a structure is called the 'Parker spiral', after a famous 1958 paper by Parker, 'Dynamics of the Interplanetary Gas and Magnetic Fields'. The spiral is readily seen in the heliospheric current sheet that separates the northward- and southward-pointing fields in Figure 34. This structure plays a role in interactions between the solar wind and Earth's protective magnetosphere, and the magnetospheres of Jupiter and Saturn. The loss of angular momentum associated with the outflow depends, as one might expect, on the stiffness of the magnetic field. This angular momentum that is lost to space exerts a torque on the Sun, slowing rotation as inferred through the solar-stellar Skumanich spin-down.

When the slow and fast wind components lie adjacent to each other in space, the spiral often hosts turbulent 'co-rotating interaction regions', formed because fast moving streams overtake slower streams. The 'Kelvin-Helmholtz' instability responsible for the interaction is also seen in regular puffs along terrestrial clouds situated in regions where winds of different speeds and/or directions overlie one another. In the solar wind, these sheared velocity fields develop into turbulence. In turn, turbulence generates wave motions and accelerates particles, which affect geomagnetic activity. The net effect is a twenty-seven-day modulation of the geomagnetic environment as these regions sweep over the Earth.

The corona, the dynamo, and CMEs

The tenuous corona may play an active role in enabling the solar dynamo to operate *at all*, as it appears to be the only agent capable of shedding a topological quantity generated by the α-effect, the magnetic helicity. The problem is that once generated, this quantity is *almost* conserved in a large and highly conducting system like the Sun. This was demonstrated by physicist John Brian Taylor (1928–) in 1974, in his studies of the relaxation of laboratory plasmas. Magnetic helicity threading solar plasma accumulates in time, even in the presence of magnetic reconnection. As a pseudo-vector, this helicity is the same for a given hemisphere (because of the sense of the Coriolis force) and a different sign in each hemisphere, independent of the sign of the poloidal magnetic fields. In other words, the magnetic helicity generated by known dynamo mechanisms must always increase in each hemisphere. Then, given Alfvén's theorem, eventually this accumulated helicity would enter implicitly the equations of motion for the plasma, and the Sun's internal motions would be forced to stop, because the equations of motion and frozen topology impose two entirely separate conditions on the system. This result is, of course, in contradiction to observations. Thus, the *Sun must find a way to shed this accumulated magnetic helicity.*

The only way out of this dilemma is for the Sun to eject, bodily, magnetic helicity. Without this ejection, the solar dynamo would stop as the twisted fields would accumulate so much stress that, like over-twisted elastic bands, they would strongly resist any more twisting, or break. The situation, seemingly unlikely, is perhaps similar to a situation where mere cloth can inflict damage on a metal machine. A twisted clump of clothes in a washing machine can break the machine through an imbalance caused by the tangling. The conjecture made explicit by B.C. Low and colleagues is that some, perhaps even the majority, of CMEs result from the need to eject magnetic helicity.

There is considerable evidence that this takes place in observations from the space age, augmenting much earlier data of cool prominence material. When we observe the plasma tracing out magnetic fields above sunspots in X-rays, we see evidence that *some flux ropes usually twist and writhe in the same direction in each hemisphere, and in different directions between hemispheres,* independent of the polarity of preceding and following spots. Coronal observations seem to reveal to us the 'handed-ness' of

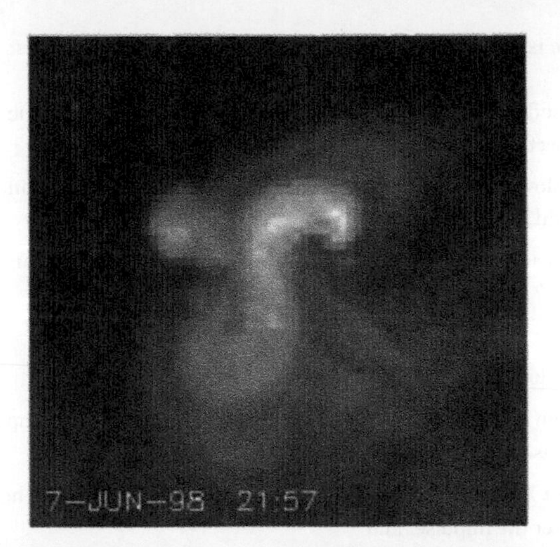

35. A soft X-ray image of an active region (08232) located at a solar latitude of -20 degrees, showing an 'S'-shaped structure, strongly suggesting that a writhing twisted rope of magnetic flux has emerged into the corona. This particular structure has a forward 'S'-shaped sigmoid typical of structures of positive helicity, seen mostly in the south solar hemisphere. The field of view covers about 1% of the visible surface.

twist in underlying sunspots, and it is usually in the sense implied by the Coriolis force. An example is shown in Figure 35, an image acquired using the Japanese Yohkoh spacecraft. This structure evolved over a few hours, ejecting plasma through the process of magnetic reconnection (change of topology) along with some of the structure's magnetic helicity.

Flares

Often associated with CMEs, flares represent yet another remarkable solar/stellar phenomenon that is only partly understood. Flares preceded CMEs in the catalogue of known solar phenomena by over a century. But in terms of physical understanding, the two appear to be intrinsically linked.

A flare is characterized by some of the following properties:

- a sudden burst of radiation over the entire electromagnetic spectrum, on time scales of minutes, the 'impulsive phase';
- a slower subsequent release of radiation in softer X-rays and UV light;
- release of an energy of about 10^{20} Joules; a major flare can shed 10^{25} Joules;
- sometimes being visible even against the photosphere;
- periodic variations in radio emission;
- high energy radiation and particle emission during the impulsive phase, extending into space;
- soft X-ray and EUV emission which can persist for many hours after the impulse; and
- frequent association with plasmoid and CME ejections.

By Earthly standards, flares emit enormous energies. The 1954 Bikini Island nuclear detonation by the United States released about 10^{16} Joules. The energy associated in each global oceanic tide is of order 10^{17} Joules. Yet all but the brightest flares are difficult to see against the Sun's disc. This is not the case for very active stars.

For a century, various imaginative causes of solar flares were debated. In 1960, Thomas Gold (1920–2004) and stalwart Fred Hoyle made clear what was involved:

> The requirements of the theory can therefore be stated quite definitely. Magnetic field configurations must be found that are capable of storing energy densities hundreds of times greater than occur in any other form, and that are stable most of the time. A situation that occurs only a small fraction of the time must be able to lead to instability in which this energy can rapidly be dissipated into heat and mass motion . . . The ultimate energy source for any

large manifestation of free energy in the solar atmosphere must no doubt be sought in the hydrodynamics of convective motion in and below the photosphere... which can result in a gradual buildup of chromospheric [and coronal electric] currents.

Alas, few more specific statements about the nature of flare energy release can be made even today; again this is a topic of active research. The questions remain open in part because flares are intrinsically unpredictable in space and time. Often the right instruments are observing the wrong part of the Sun at the right time, and vice versa. But observations include remarkable full-Sun imaging at high energy X- and γ-ray wavelengths from the RHESSI satellite. Imaging at high energies is achieved using ingenious technology based on time-modulated signals from a spacecraft spinning fifteen times a minute. High energy radiation can only be generated efficiently by particles with high energy. The generation of high energy particles requires macroscopic electric fields, and so this naturally lies beyond MHD, which cannot sustain any electric field in the frame of the moving plasma. To a physicist this is exciting. It points to physics beyond the fluid approximation, implying that high frequency plasma phenomena (such as turbulence or collective plasma effects) are important.

At present the conditions under which flares occur remain unknown. Observations have revealed configurations that seem to be associated with flares. The emergence of a twisted flux rope through the atmosphere and into the corona has been inferred both by photospheric Zeeman Effect work, and through the regular association of flares with 'sigmoids'—twisted, writhing shapes in coronal loops in soft X-ray data (Figure 35). Such work continues to be aggressively pursued.

Figure 36 shows the 'standard model' for solar flares. The model is a qualitative picture, a kind of hypothesis to be tested and perhaps refuted by critical observations. The figure shows a two-dimensional cut through a three-dimensional arcade of

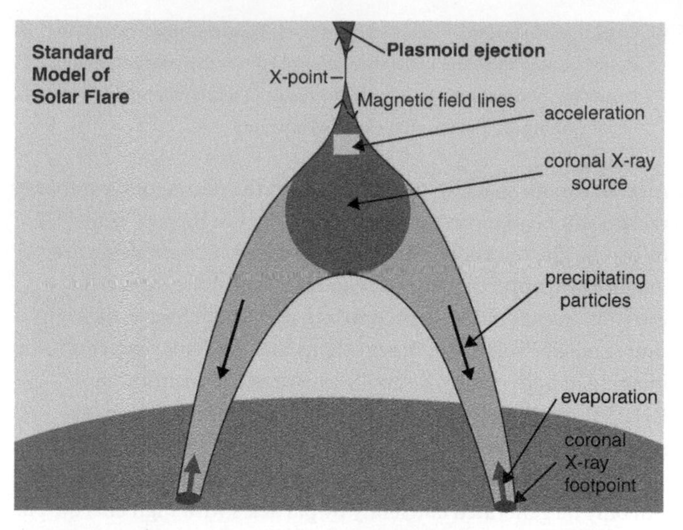

36. The 'standard model' of solar flares. The figure shows events following the disruption of a loop in an active region (see, e.g., the loops in Figure 30), by the upward eruption of part of the loop (labelled the 'plasmoid'). The magnetic field lines are brought together under the plasmoid at the X-point, and these begin to reconnect, freeing the field lines below to join the underlying loop and those field lines above to be let loose, embedded in the plasmoid, forming a CME.

coronal loops. Loss of equilibrium (because of excess free energy stored in the magnetic field) leads to the ejection of a plasmoid, which, in some flares, will be a CME. This ejection leads to the 'X-point' configuration and current sheet underneath, suggesting the onset of rapid reconnection (Figure 27). A variety of instabilities drawing on free magnetic energy can lead to ejection.

Until we can really measure the changing magnetic field within the corona during flares, we can only conjecture as to how the processes such as acceleration, energy, and particle transport happen. For example, no data definitively show that high energy particles are actually directed towards the solar surface as in the standard model. Instead, we have evidence that these fast, non-thermal particles merely exist, but with no directional information.

In summary, current theoretical ideas suggest that CMEs can drive flares, but CMEs may not be necessary or sufficient. Our ideas suggest that CMEs must exist in order to rid the Sun of certain quantities, such as magnetic helicity, generated by solar rotation, and intrinsic to the solar dynamo. Perhaps this sink of magnetic helicity is as important to the Sun as a sink of digested food is for mammals.

A surprising role for solar magnetism

In closing this chapter, it is worth pointing to remarkable features about the Sun that were appreciated, perhaps before their time, by luminaries such as Richard N. Thomas (1921–96) and R. Grant Athay. Thomas was a fearsome advocate of seeing the big picture, famous for his intimidating style in meetings, often accompanied by his colleague Anne Underhill (1920–2003). Athay, less confrontational, for years held a global view of the Sun, summarized in his last paper, a 2008 work with his colleagues Low and White. The Sun and stars are examples of systems that are *open to space*. In losing mass, angular momentum, energy, magnetic helicity, and other quantities, the Sun is a system with intricate and delicate feedbacks. Even though the energy associated with its unanticipated varying magnetic field is weak, it lies at the heart of these couplings. No-one could have easily anticipated that a dynamo might be intrinsically tied to a magnetic helicity that must be shed bodily from the Sun, for a cyclic dynamo to operate. To a physicist, this is an eye-opening perspective. Such non-intuitive surprises are a sign that solar physics will continue to be a profitable research area for the foreseeable future.

Chapter 5
Solar impacts on Earth

The Earth is subject to influences from space, and the largest influences arise from the Sun. A large fraction of the Sun's radiation reaches the ground and ocean, but by no means all. In earlier chapters we saw that the Sun's radiant variations summed over all wavelengths are modest, measured changes being a bit less than about 0.1 per cent over recent decades. The net effect of the bulk of the Sun's radiation is to warm our planet. This situation is very conducive to life.

But over far-UV and X-ray wavelengths, the decadal changes can reach factors of 1.5 to 10, increasing with the energy of the radiation, as the sunspots come and go. This UV and X-ray radiation is absorbed in the high atmosphere, forming the thermosphere and ionosphere above about 80 km altitude. Historically, this region has two names, because heat is added and ions are created, through the process of photo ionization by the UV and X-radiation. The ionosphere is the part of the thermosphere that is substantially ionized. It has a daily cycle as the Earth rotates relative to the Sun's radiation and our atmosphere adjusts to this influence. A cartoon sketch of our atmosphere is reproduced in Figure 37.

The near-UV radiation (200 nm to 315 nm wavelength) is absorbed by ozone much lower in the stratosphere, from about 15

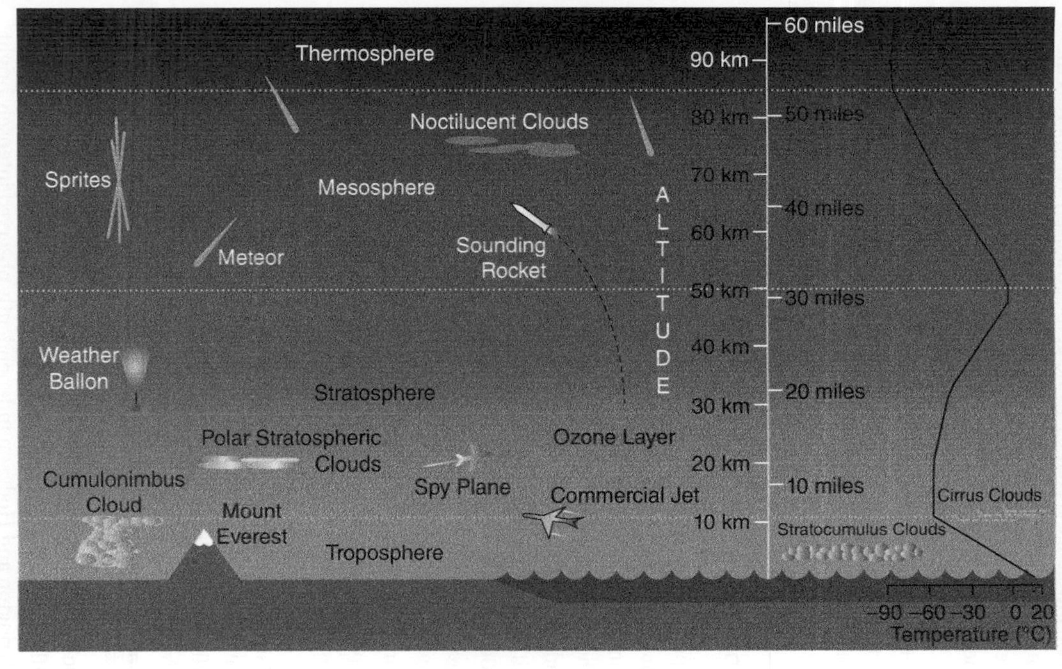

37. Earth's atmosphere.

to 30 km altitude. Curiously, solar radiation at the shorter end of these wavelengths creates ozone in the stratosphere, whereas the longer wavelength radiation destroys it. Between stratosphere and thermosphere/ionosphere is the coldest region called the 'mesosphere'.

The coupling between these layers of the atmosphere, modulated by large solar magnetic variations, is a subject of intense research. Small effects of sunspot variations can even be traced even down to the Earth's surface.

Space weather

'Space weather'—the rapidly changing conditions in the Earth's ionosphere and above—has become a topic of military, commercial, political as well as academic interest. The reason? The natural tendency for the Sun to eject variable high energy radiation, particles, and magnetic fields presents modern society with problems that were not relevant to our ancestors. For the 300,000 or so years of the emergence of *Homo sapiens*, our geological and other records reveal that the Sun has behaved un-spectacularly, and it continues to do so. What *has* changed is our technology, our increasing dependence on electronics, especially those in orbit around the Earth, and in our worldwide power supply networks. Almost all space assets, including the GPS system, lie between low and geostationary Earth orbits, well inside our protective 'magnetosphere'. But this magnetosphere is not entirely stable, and can be made open to solar and other influences under the right (or perhaps wrong) conditions. The effects of weather in space can be illustrated by historical examples.

One solar magnetic eruption, 'The Carrington Event' in 1859, even influenced the electrical infrastructure of the day, crude as it was. Carrington and Hodgson independently observed intensely bright ribbons, seen among groups of sunspots on 1 September 1859. The emission originated from a flare associated with the release of what we know now as a CME, that travelled the 150 million km towards Earth in eighteen hours, at 0.8 per cent of the speed of light. This CME opened up Earth's protective magnetosphere, allowing solar charged particles into the atmosphere. The Sun does not care if its CME magnetic fields align with magnetospheric fields. So again, tangentially discontinuous magnetic fields are brought together (Figure 27) enabling rapid reconnection, and allowing solar plasma to enter Earth's atmosphere. There followed spectacular auroral activity and large fluctuations in the Earth's surface magnetic field. The varying magnetic field, by Faraday's law of electromagnetic induction, spawned electric fields and currents along conductors on the Earth's surface. The effects were remarkable. On the ground, a conversation between telegraph operators was reported in *The Boston Globe*:

BOSTON: *Please cut off your battery entirely for fifteen minutes.*
PORTLAND: *Will do so. It is now disconnected.*

BOSTON: *Mine is disconnected, and we are working with the auroral current. How do you receive my writing?*
PORTLAND: *Better than with our batteries on . . .*

This event occurred during a sunspot cycle of high amplitude (Figure 19), but several cycles have been more intense.

In a second example, on 13 March 1989, a similar solar-generated 'geomagnetic storm' took down Hydro-Québec's supply of power to millions for several hours. On 6 March, a very large solar flare was observed. A large CME was detected heading towards Earth on 10 March. Subsequently, by the 13th, a severe geomagnetic storm began with intense aurorae which could be seen as far south as Florida. The Québec power was interrupted, power grids were affected as far away as New York, and several satellites suffered significant glitches, some sent into a temporary tumble. In 2012 the Earth dodged a more energetic CME.

The first records of space weather lie in cave paintings of aurorae, suspected to have been depicted as long ago as 30,000 BCE by Cro-Magnon people in southern France. Written auroral records can be found as far back as 2600 BCE in China. Galileo Galilei named the phenomenon 'Aurora Borealis' in 1619, after Aurora, the Greek goddess of the morning, presuming that the phenomenon was akin to dawn. Quantitative measurements of aurorae began in the late 18th century. Cavendish used triangulation to determine that the phenomenon lay near 100 km above the Earth's surface.

The first hints of a solar origin for aurorae were seen in data in the 19th century, beautifully documented in C.A. Young's book *The Sun*, several editions of which were published before 1900. (Copies of the book can readily be found today.) Direct evidence that the Sun was responsible was scant but tantalizing, the Carrington Event being highly suggestive. Evidence mounted for an association between geomagnetic disturbances and the twenty-seven-day solar rotation. By 1904, Edward Maunder quantified relationships between sunspots and geomagnetic activity, uncovering a significant correlation between geomagnetic disturbances and solar rotation. He concluded that

> our magnetic disturbances have their origin in the Sun. The solar action which gives rise to them does not act equally in all directions, but along narrow, well defined streams, not necessarily radial.

In the 1930s Julius Bartels (1899–1964) had related these disturbances to what he called solar M regions. It took another four decades for these mysterious regions to be identified. During the 1973–4 SKYLAB era, sufficient data were acquired to prove that dark coronal regions—coronal holes—were related to the hypothetical 'M regions', nicely tying together work begun by

Maunder and others. Credit for the first association goes to Carole Jordan (1941–) in 1974. Coronal holes were identified as the sources for the 'fast' solar wind.

Space weather assumed worldwide political importance with the advent of World War Two, as both Axis and Allied powers found themselves periodically unable to communicate with military assets, among other problems. Two observatories were set up on either side of the Atlantic. The Axis powers, under the guidance of Karl-Otto Kiepenheuer (1910–75), built a network of solar observatories across Europe. The Kiepenheuer Institute for Solar Physics in Freiburg was named in his honour. In the USA, Donald Menzel (1901–76) sent Walter Orr Roberts (1915–90) to Climax, Colorado (elevation 3,465 metres) to observe the solar corona and prominences using a Lyot coronagraph. Thus began the original HAO, now part of the National Center for Atmospheric Research in Boulder, Colorado. Today there are several organizations worldwide, with observatories on Earth and in space, devoted to space weather.

Under 'normal' conditions the Sun has a relatively benign influence on space near the Earth, because our magnetosphere serves as a barrier around which, like water flowing past the bow of a ship, the solar wind simply flows past us. But during storm conditions, the ship's hull is breached! The dominant physical processes behind geomagnetic storms like that which occurred in 1859 are summarized in Figure 38. The configuration shown is that of a geomagnetic storm; the solar magnetic field has the opposite polarity to that of the Earth at the region between the Sun and Earth called the 'magnetopause', forming the familiar 'X-point' in the magnetic field (Figures 27 and 36). Magnetic reconnection is enforced in both Sun and anti-Sun directions. Both regions lead to aurorae and geomagnetic electric currents.

In the simplest physical description, the dynamical reactions are dominated by one parameter, which is the direction of the solar

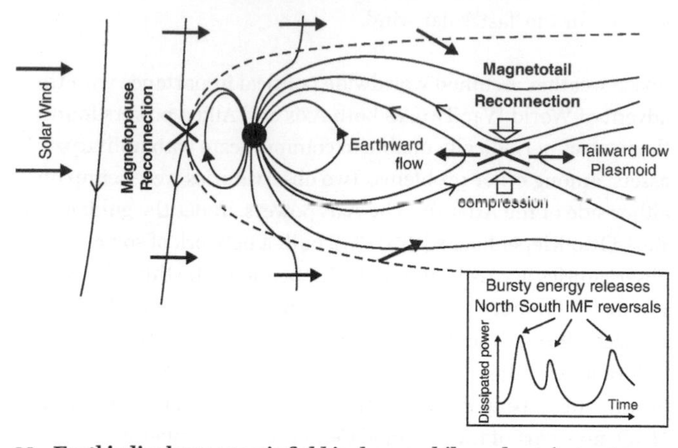

38. Earth's dipolar magnetic field is shown while undergoing a strong perturbation from the solar wind or a CME. The configuration is that of a geomagnetic storm.

magnetic field relative to that of the Earth's field, at the magnetopause (Figure 38). When they are mostly aligned, magnetic reconnection will be weak and the topology mostly unchanged. But when the fields are oppositely oriented, reconnection is likely to change topology rapidly, admitting solar plasma into the Earth's environment. The reconnection can occur on both night and day sides of the Earth, which permits energetic solar particles to enter the otherwise protective magnetosphere which normally simply diverts solar wind flow around the Earth, the magnetosphere suffering a small compression on the day side. The detailed (microscopic) physics of this reconnection differs a little from the solar case in that the densities in the magnetosphere are orders of magnitude below those in the corona. The 'fluid' picture breaks down and the dynamic evolution of the reconnection depends on so-called collisionless processes.

Curiously, research suggests that the rates of reconnection may not depend much on the small-scale physics. Once a large-scale configuration susceptible to reconnection has arisen, it will evolve in a fashion to make fast reconnection happen, independent of the micro-physics. This situation reflects a (partially understood) coupling between physically small and large scales.

Climate

Homo sapiens arose at the latest moment in the life of the solar system, just the final six seconds, if we count the whole history of the solar system as one day. Aeons were needed for life on Earth to take hold and generate significant free oxygen in the atmosphere, permitting multi-cellular life to arise. The first complex cells, or eucaryotes, are believed to have arisen about 1.5 thousand million years ago, later than generally believed prior to about 2015. The first sexual reproduction occurred about one thousand million years ago. During these periods, the Sun must have influenced the Earth in a benign fashion. Changes of even a few per cent in the amount of radiation reaching the Earth, or occasional large disruptions of our magnetosphere, would have been sufficient to alter significantly the evolution of life, even halt the onward march of complexity of life forms. Our mere existence tells us something about the stability of the Sun and the solar system that is quite profound, for, as we saw in Chapter 2, the main-sequence Sun, as stars go, is a rather dull and mildly variable object. Some readers may feel concerned if, while reading these words, they recall some of the Sun's early history summarized in Chapter 2. Although the main sequence is stable over many aeons, the Sun's luminosity certainly increased by tens of per cent while it burned nuclear fuel early on the main sequence. This is hardly a constant supply of solar energy, albeit over a very long time. The newly formed Earth was illuminated by only 70 per cent of today's radiation. Yet, there is evidence of liquid water on the Earth's surface soon after the Earth formed, after just one hundred million years. This leads us to the famous 'Faint Young Sun paradox', *how was the Earth (perhaps even Mars) warm enough to permit liquid water on its surface, yet receiving just over two-thirds of today's solar radiation flux?* Astronomers, biologists, geophysicists, palaeontologists, and meteorologists all have something to offer in this modern field of research.

The 'Faint Young Sun' paradox

The problem was raised by Carl Sagan (1934–96) and George Mullen in 1972. Today, the radiant energy received at Earth from the Sun is about 1,361 W/m^2. Between maximum and minimum sunspot numbers, the extremes are a little less than 1 W/m^2. As the Sun joined the main sequence 4.5 thousand million years ago, the radiation would have been a mere 950 W/m^2. If the Earth radiated perfectly, then the amount of energy radiated is the surface temperature to the fourth power ('Stefan's law' after Josef Stefan, 1835–93). Ludwig Boltzmann derived this relationship from first principles in 1884. In the absence of other heating mechanisms (such as radioactive decay inside the Earth, or lunar tidal heating), the temperature of the Earth's surface would be determined by the balance of the outgoing and incoming radiation. Then the surface temperature of the Earth as the Sun began life on the main sequence would be just 91 per cent of today's average of 284 K, that is, 263 K. While this may seem a modest change, this is 10 K below the freezing point of water.

Thus, the early Earth should have been a snowball. This contradicts a plethora of evidence from geology, and biology, documented in records imprinted in rocks and fossils. For cellular life to appear, liquid water must be present as a substrate for the cells to function. Geothermal activity, driven by radioactive decay in the Earth and tectonic activity could have locally produced life forms on the early Earth. But under a solid frozen umbrella, it could not have spread globally. The geological record reveals the Earth's surface to have been continuously warm (with a period called the Huronian glaciation period, 2.4 to 2.1 thousand million years ago, and a later 'Snowball Earth' period between 850 and 635 million years ago). Surface water-related sediments have been dated at 3.8 thousand million years ago.

Several resolutions to the paradox have been proposed, falling into conjectures that are either astrophysical or terrestrial. The following are astrophysical in nature:

- The Sun was significantly more massive, hence brighter, when it joined the main sequence. It has lost mass during its evolution sufficient to explain the problem.
- The Earth's surface was heated by stronger tidal forces from our Moon orbiting closer to the Earth than today.
- The Sun was spinning faster, and was more magnetically active when young, enhancing the total energy through very bright plage regions in the sunspot zones, preferentially sending energy into the plane of the Earth's orbit.
- Galactic cosmic rays are known to be modulated by the solar magnetic field. Perhaps the early Sun's magnetic fields changed this flux, and thereby altered the rates of cloud formation, and the amount of energy reflected into space, in the early Earth. (Clouds form when free water molecules condense into droplets in the air. 'Cloud chambers' used in sub-atomic physics uses this same effect to track ionized particles as the charged particles cause the water molecules to coagulate along the particle tracks.)

In the opinion of the author, though even the first possibility is perhaps unlikely, the list is in decreasing order of likelihood. Stellar observations have been used to estimate mass loss rates of young Sun-like stars. At a given stage of evolution, the luminosity of a main sequence star is proportional to roughly the fifth power of the mass (this is a well-established result of stellar evolution theory called the 'mass-luminosity relation'). So we could resolve the paradox if the Sun were 'just' 7 per cent more massive upon reaching the main sequence. The Sun currently loses at most three parts in 10^{14} of its mass per year. In the first thousand million years of its existence, the Sun would have had to have lost about seven parts in 10^{11} per year, for the paradox to be resolved. Work by astronomer Jeffrey Linsky (1942–) and colleagues suggested that the Sun could have lost at most 3 per cent of its mass since arriving on the main sequence. Very recent work by Skumanich suggests that this is a strict upper limit. Skumanich's work suggests a much weaker dependence of mass loss on age. The implantation of solar wind ions on the surfaces of meteorites also suggests that a high solar wind flux existed for too short a time, about a hundred million years, to be entirely consistent with the higher estimate.

Still, these astronomical solutions are attractive in that some may also explain the evidence for the presence of water on early Mars. Two planet records are better than one. For the terrestrial propositions, we have

- Earth could have undergone significant greenhouse warming, because of a very different chemical balance in the early atmosphere.
- Estimates of radioactive heating in the past are small (perhaps 0.1% of solar). However, the extra internal heat could have changed surface out-gassing conditions, augmenting the greenhouse effect.

Given that Venus has undergone catastrophic global warming, the first point is also appealing.

The problem, still unresolved, remains of great interest, in particular with the growing plethora of extrasolar planets discovered and the continuing quest to understand the origins of life.

The Sun and Earth's climate

Early attempts to measure changes in the Sun's irradiance of Earth were begun from the ground early in the 20th century. Charles Greeley Abbot (1872–1973) expended considerable energy trying to measure solar brightness variations. Sadly, satellite measurements during the space age revealed that these were doomed to fail because the variations lie well below levels of variation detectable from the ground.

Measurements of irradiance since 1978 have been made from satellites which have a far higher precision. We know now that the radiation flux varies by about 0.1 per cent from sunspot minimum to maximum, because some satellite experiments have lived long enough to sample the half-cycle of eleven years, and their calibration is believed to be precise, but with a lower accuracy. (A precise measurement, one where the results are reproducible within a very small range, can nevertheless be quite inaccurate, and far from the 'true' value of the physical parameter.) Measurements made with different satellite experiments disagree by as much as 0.4 per cent. The result is that we know the amount that the Sun varies in its radiative flux at Earth, but *only on a time scale of about ten years*.

For stars we have a far longer (over twenty-five-year) database of the *visible* brightness changes of Sun-like stars with a precision close to 0.1 per cent, thanks to the efforts of stalwart astronomers such as Greg W. Henry (1950–) and Louis J. Boyd (1945–). Sun-like stars tend to vary in a similar fashion to the Sun, but the Sun may be a little less variable than most of its cousins.

When it comes to time scales of a century, 1,000, or 10,000 years (the holocene), we are on less solid ground. In fact, we are led to mere extrapolation in our attempts to understand solar radiative variations. In other words, we are forced to *assume* a behaviour of the Sun, based upon some data obtained over recent years, and then extrapolate into the past. As any follower of the stock market knows, such procedures are risky. Any kind of extrapolation must be accepted as an educated guess.

The kind of extrapolation schemes adopted are as follows. First, seek a feature (such as sunspots, tree rings, or ice cores) which records some factor that is related to the Sun's activity. Next, examine variations in solar irradiance in the past few decades in which direct measurements are available. Draw a graph of terrestrial and solar properties against the irradiance. If one follows the other, then we can postulate that this behaviour continued throughout history. Then, even though we have solar measurements only since 1610 (sunspots) or far later (Galactic cosmic rays), we can use tree rings and ice cores for thousands of years to estimate the solar quantities in the past. Such works have been called 'reconstructions'. Sometimes, reconstructions become much more sophisticated by including some kind of physical model or process to augment the data. Such procedures do not, however, imply accuracy.

The problem here should be self-evident. There is generally no way to test such reconstructions. They might be pitted against data from other sources and inconsistencies sought (you cannot have a negative number of sunspots, for example), but the Sun's history will forever remain out of the realm of measurement. As such, Popper would classify these extrapolations as 'pseudo-science'.

Unlike such extrapolations, Jack Eddy's historical survey of sunspots and geomagnetic activity extended back to around 1610 when observers such as Galileo were able to use the first telescopes. The startling conclusion was that, during what he dubbed the Maunder Minimum (see Figure 19), there was a severe dearth of sunspots and accompanying aurorae. Intense scrutiny since has served to confirm Eddy's overall conclusions. Now Eddy took a further step to suggest that weather might be connected to this seventy-year solar phenomenon. Written history records a drop in temperature across Europe over decades surrounding this period, and there is geological evidence that a reduced temperature was experienced worldwide. The Little Ice Age extended far longer than the Maunder Minimum. But the deepest freeze of the Little Ice Age roughly coincided with it. Unfortunately, recent attempts to reconstruct the radiative flux from the Sun made through this period differ widely in their results. Large discrepancies are perhaps expected, such is the nature of extrapolations. But based upon Eddy's and related analyses, the question of the relationship between climate and sunspots remains interesting.

The scientific community does not yet agree on the link between the eleven-year variation in sunspot numbers and terrestrial weather from the data alone. At least part of the problem is that weather is intrinsically local and highly variable, so that it is difficult to identify very small decadal changes in the strongly-varying signals. Further, the oceans respond to changes in heat on time scales also of a decade, and the net signals of around a decade that we try to interpret can be confusing.

Although this book is about solar physics, it would be incomplete without a statement of the effects of the Sun on the modern trend in global warming of the terrestrial atmosphere. Again we face the problem of having only accurate short-term measurements of the solar radiation incident on the Earth. But the totality of evidence from the Sun's measured behaviour, the physics behind modulation of the radiative flux, as well as the behaviour of stars, strongly indicates that there has been no heating trend in the Sun that correlates in any way with the recent global warming trend. This conclusion is supported by detailed simulations of climate changes driven by various assumptions concerning solar variability, such as are done at the National Center for Atmospheric Research in Boulder, Colorado. Comparisons of extreme variations over periods of decades to centuries with control numerical experiments simply cannot be used to attribute recent global warming to solar activity. Finally, recent cycles of sunspots are quantitatively similar to those experienced around 1900, when the average temperature of the Earth was about 1 K less than today, according to National Oceanic and Atmospheric Administration (NOAA). Thus the recent 'deep minimum' in sunspot activity of 2008 appears to have been preceded by earlier, similar solar episodes, in which there was no changed climate, in stark contrast to the recent and unprecedented increase in global warming. All evidence suggests that the recent global warming is due to human activity.

Chapter 6
The future of solar physics

The future of solar research seems to be intimately tied to the effects the variable Sun has on the Earth. Solar research will grow as several truly novel facilities online now, or soon to come online, make their impact. Perhaps first among these is the DKIST, which is on the mountain Haleakala on the US Island of Maui, Hawaii (see Figure 39).

Unlike most earlier ground-based telescopes, the DKIST has a definite mission, akin perhaps to missions to Mars and other planets. DKIST was designed and built explicitly to answer how the magnetic field of the Sun evolves at its surface, and within its corona. It has a mirror of diameter 4 metres, with a huge collecting area six times larger than any previous solar telescope, and a focal length of just 8 metres. This 'focal ratio' of f/2 will be understood by photographers to be a very 'fast' system, like an 'f/2' camera lens, or a magnifying glass. In consequence, at the main focus it must deal with a heat load of up to 3 mega-W/m^2. This requires special engineering at the main focus, sending as much of the reflected radiation back into space as possible.

But this is just the first of DKIST's unique qualities. It has a super-smooth main mirror that reflects light from the atmospheric UV cut-off near 310 nm, to 28 microns, a far infrared wavelength. The wavelength range spans almost a factor of a hundred. This

39. DKIST enclosure, situated atop Haleakala mountain on the Hawaiian island of Maui.

mirror is deliberately placed 'off-axis', so that pure sunlight, unimpeded by any part of the telescope, hits the main mirror and is reflected upwards and a little sideways. The mirror can be thought of as a circular side section of a much larger mirror of a 'normal', symmetrical telescope. It must be kept extremely clean, and the telescope enclosure is especially designed to optimize the steady movement of air throughout the observatory. This avoids small pockets of convection that can otherwise build up over hot-spots, to help keep the images steady. With these properties, it is also designed to measure the very dim light of the solar corona adjacent to the bright photosphere, and the darkest regions of sunspots. The technology and design is so new, the DKIST started life as the 'Advanced Technology Solar Telescope', before being named after a prominent advocate for the project, US Senator Daniel K. Inouye. Current and future scientists will use DKIST's sophisticated new instruments placed on a 16-metre-wide rotating table under the telescope, to study the evolving solar magnetism in exquisite detail. We hope to see, through measurements of coronal magnetic fields and plasma, just how the Sun's magnetic fields generate flares, CMEs, and the solar wind, providing long-sought answers to pressing yet old questions outlined in this book.

Other major missions include NASA's Parker Solar Probe, launched in 2018, and the European Solar Orbiter mission, successfully launched in February 2020. As their names imply, these spacecraft will orbit the Sun in novel ways, exploring plasma deep within the solar wind and observing the Sun not only *in situ*, but also from very different viewpoints from Earth. The Parker probe is the fastest object ever made by humankind: it will achieve speeds of 690,000 km per hour. At this speed the probe will traverse 1 solar radius in one hour! This speed is achieved partly by rockets that send the probe out of Earth's gravity, partly by helpful pulls by Venus fly-bys, and mostly by the pull of the Sun's gravity as the probe ultimately approaches to within 10 solar radii in 2025. The radiant heating experienced then is 1 per cent of that experienced by plasma at the solar surface, not the 1 kW/m^2 we feel at Earth, but 600 kW/m^2. While smaller than the heat load of DKIST, the probe must nevertheless be carefully protected. The probe therefore has a novel heat shield behind which the spacecraft and instruments shelter. This can be seen on the top of the spacecraft in Figure 40.

40. The Parker Solar Probe is seen being hoisted on to the third stage rocket motor on 11 July 2018.

The *in situ* measurements provide *complementary* views of processes within solar magnetized plasmas. Ground-based observatories can see, for example, flares on the Sun, and the probes can measure their effects as the mass, momentum, and energy passes into space. DKIST observations, perhaps with large new ground-based telescopes being developed in Europe and China, will permit us to look directly at coronal magnetic fields on scales above, say, a few thousand kilometres. But these new spacecraft can probe magnetic fields and the entrained plasma at scales down to metres. Using these new instruments in combination, we might expect to solve the puzzle of coronal heating, study the nature of turbulence in magnetized plasmas, and understand how the Sun sheds the accumulated magnetic helicity in order to permit the entire solar magnetic engine to sustain its remarkable twenty-two-year magnetic cycle.

These missions, supported by the ever-increasing power of computers, are ushering in a new era. There has never been a more exciting time to be studying our neighbour, the Sun.

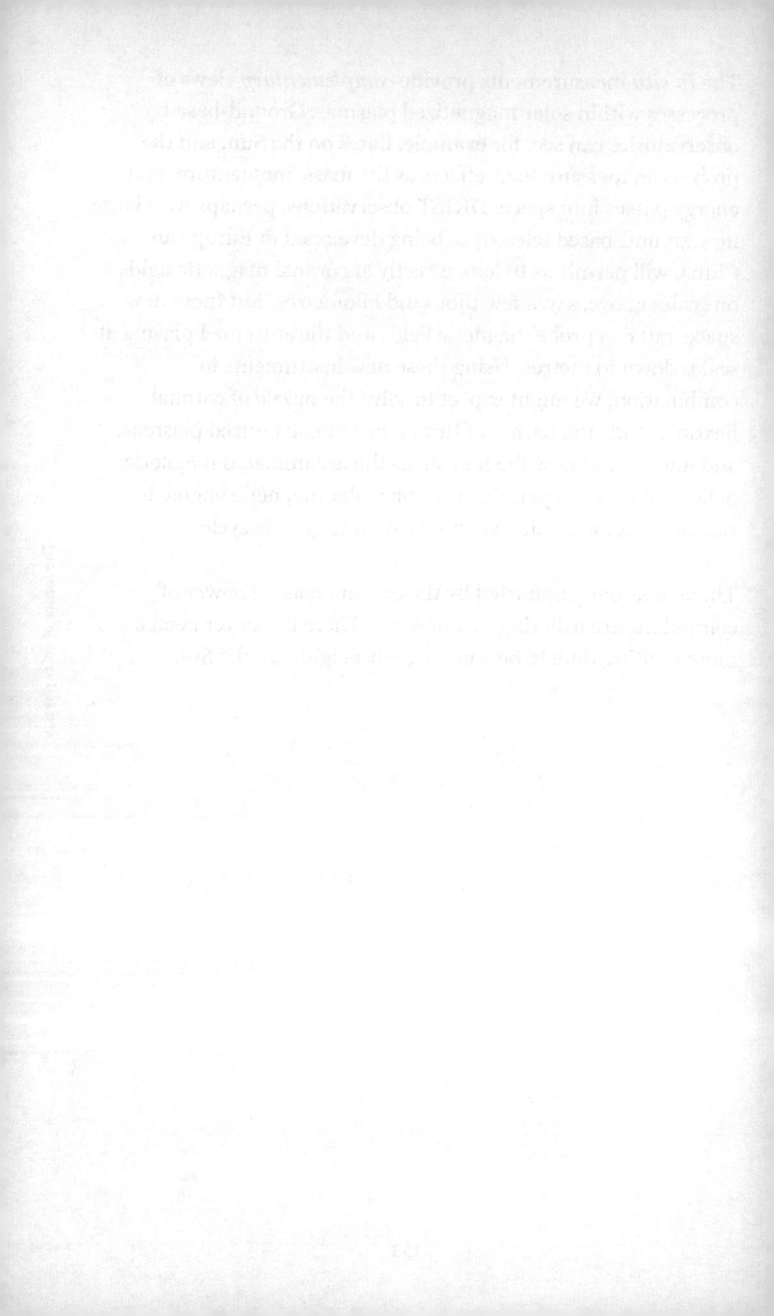

Further reading

Chapter 1: The Sun, our star

Young, C.A., *The Sun*, 4th edn. Kegan Paul, Trench, Truebner & Co. Ltd., 1892. This book is still available in many used book stores; readers will be rewarded with much insight into 19th-century knowledge of solar and solar-terrestrial physics.

Green, L., *15 Million Degrees: A Journey to the Centre of the Sun*. Viking, 2016. This is a very readable, enthusiastic book, including an acknowledgement of the role of women in our exploration of solar physics.

Foukal, P., *Solar Astrophysics*, 3rd edn. Wiley-VCF, 2010. This book introduces solar observations and theory, including an introduction to space weather, at a level suitable for those with a technical background interested in understanding more deeply solar physics.

Chapter 2: The Sun's life-cycle

Gamow, G., *The Birth and Death of the Sun: Stellar Evolution and Subatomic Energy*. Dover, 2005. A reprinting of a 1952 version. The book pre-dates the discovery of the processes leading to elements heavier than iron, but is interesting for the general reader.

King, A., *Stars: A Very Short Introduction*. Oxford University Press, 2012. A summary of the lives of stars of all masses, including the evolution to black holes.

Tayler, R., *The Stars: Their Structure and Evolution*. Cambridge University Press, 2010. This is a more technical introduction using physics and mathematics at about GCSE level.

Chapter 3: Spots and magnetic fields

Schatz, D. and Fraknoi, A., *Solar Science: Exploring Sunspots, Seasons, Eclipses, and More*. NSTA Press, 2016. This is a book intended for teachers and children from 10 years and up, available also as an electronic edition.

Parker, E., *Conversations on Electric and Magnetic Fields in the Cosmos*. Princeton Series in Astrophysics. Princeton University Press, 2007. This short book is a highly recommended collection of thoughts about how space plasmas work based upon decades of experience and insight, for technically-minded people with degrees in technical subjects.

Cowling, T.G., *Magnetohydrodynamics*. Hilger, 1976. The author presents a short introduction to magnetohydrodynamics at technical/science degree level.

Zirker, J., *Sunquakes: Probing the Interior of the Sun*. Johns Hopkins University Press, 2003. The monograph presents a very readable account of helioseismology.

Chapter 4: The dynamic corona

Zirker, J., *Total Eclipses of the Sun*. Princeton University Press, 1995. The author presents a readable story about eclipses over thirty centuries, discussing physics of the atmospheres of the Sun and Earth at a level suitable for all.

Golub, L. and Pasachoff, J.M., *The Solar Corona*. Cambridge University Press, 2nd edn, 2010. The book contains a nice discussion of coronal observations including those obtained during the 'space age', but the physics and mathematics assumes degree level knowledge in physics or a related subject.

Chapter 5: Solar impacts on Earth

Eddy, J., *The Sun, the Earth, and Near-Earth Space: A Guide to the Sun-Earth System*. NASA, 2017. Suitable for all, the book provides an enjoyable discussion of the effects of the Sun on Earth and our technology.

Index

D

E

F

G

H

I

J

K

L

M

N

T

U

W

Y

Z

EPIDEMIOLOGY

A Very Short Introduction

Rodolfo Saracci

Epidemiology has had an impact on many areas of medicine;
and lung cancer, to the origin and spread of new epidemics.
and lung cancer, to the origin and spread of new epidemics.
However, it is often poorly understood, largely due to misrepresentations in the media. In this *Very Short Introduction* Rodolfo Saracci dispels some of the myths surrounding the study of epidemiology. He provides a general explanation of the principles behind clinical trials, and explains the nature of basic statistics concerning disease. He also looks at the ethical and political issues related to obtaining and using information concerning patients, and trials involving placebos.

www.oup.com/vsi

The History of Medicine

A Very Short Introduction

William Bynum

Against the backdrop of unprecedented concern for the future of health care, this Very Short Introduction surveys the history of medicine from classical times to the present. Focussing on the key turning points in the history of Western medicine, such as the advent of hospitals and the rise of experimental medicine, Bill Bynum offers insights into medicine's past, while at the same time engaging with contemporary issues, discoveries, and controversies.

SUPERCONDUCTIVITY

A Very Short Introduction

Stephen J. Blundell

Superconductivity is one of the most exciting areas of research in physics today. Outlining the history of its discovery, and the race to understand its many mysterious and counter-intuitive phenomena, this *Very Short Introduction* explains in accessible terms the theories that have been developed, and how they have influenced other areas of science, including the Higgs boson of particle physics and ideas about the early Universe. It is an engaging and informative account of a fascinating scientific detective story, and an intelligible insight into some deep and beautiful ideas of physics.

ONLINE CATALOGUE

A Very Short Introduction

Our online catalogue is designed to make it easy to find your ideal Very Short Introduction. View the entire collection by subject area, watch author videos, read sample chapters, and download reading guides.

http://fds.oup.com/www.oup.co.uk/general/vsi/index.html

SOCIAL MEDIA
Very Short Introduction

Join our community

www.oup.com/vsi

- Join us online at the official Very Short Introductions **Facebook** page.
- Access the thoughts and musings of our authors with our online **blog**.
- Sign up for our monthly **e-newsletter** to receive information on all new titles publishing that month.
- Browse the full range of Very Short Introductions online.
- Read **extracts** from the Introductions for free.
- Visit our library of **Reading Guides**. These guides, written by our expert authors will help you to question again, why you think what you think.
- If you are a teacher or lecturer you can order inspection copies quickly and simply via our website.